Fern Flora of South India

Taxonomic Revision of Polypodioid Ferns

Santhosh Nampy

P.V. Madhusoodanan

1998

DAYA PUBLISHING HOUSE

DELHI-110035

ISBN 81 - 7035 – 179 – 0

Published by : **Daya Publishing House**
1123/74, Deva Ram Park
Tri Nagar, Delhi-110035
Phone : 7103999
Fax: (011) 7198902

Laser Composed by : **Delite Printers**
Delhi-110035, Ph: 7182133, 7182135

Printed at : **Chawla Offset**
Delhi

PRINTED IN INDIA

ACKNOWLEDGEMENTS

The authors express their sincere gratitude to Dr. K. Unnikrishnan, Head of the Department of Botany, University of Calicut; Fr. M. T. James, Principal, St. Joseph's College, Devagiri; Dr. P. Baas, Director, Rijksherbarium, Leiden; the curators of the herbarium, Royal Botanic Gardens, Kew and British Museum Natural History, London for providing facilities; Ms. Bertie van Heuvan, Rijksherbarium for SEM photographs and to our friends and colleagues in the Department for various help during this work.

We would like to extend our sincere appreciation to Dr. M. G. Price, Dr. Dan H. Nicolson (USA), Dr. Hans P. Nooteboom, Dr. Monique Bosman (The Netherlands), Dr. B. S. Parris (New Zealand), Dr. Zhang Xian-Chun (China), Dr. V. J. Nair, Mr. Peter Edwards (U. K) and Dr. V. S. Manickam (India) who have been kind enough to help us in various ways. The services of curators of Museum National d Histoire Naturelle Laboratoire de Phanerogamie (Paris), Real Jardin Botanico (Madrid), Rijksherbarium (The Netherlands), Mr. Bhargavan and Joint Director, Botanical Survey of India (Coimbatore) are greatfully acknowledged.

We must also thank Dr. A. K. Pradeep, Dr. K. R. Leena, Dr. Benny Thomas, Mrs. Shiby Santhosh, Mr. Shaji, Dr. M. Sabu who were of great help during this work. We also express our gratitude to Mr. Anil Mittal of M/s Daya Publishing House, New Delhi for publishing the book.

Santhosh Nampy
P. V. Madhusoodanan

CONTENTS

LIST OF FIGURES

INTRODUCTION

Importance of the present work

Pteridophytes, the pioneer colonizers on land, characterised with independent alternation of generation and primitive vasculature, are very conspicuous and elegant elements of our present day flora. They are mainly confined to the tropical, subtropical, and warm-temperate regions of the world representing a wide spectrum of ecological types ranging from small filmy ferns to arborescent types and xerophytes to hydrophytes.

The distribution and abundance of ferns in India are fairly well known, and about 800 fern species are so far reported from here. Of the different phytogeographical zones of India, the Western Ghats and the Himalayas are rich in fern diversity. The Western Ghat region of South India alone is the abode of more than 250 species of ferns.The unique physiography, mountainous terrain with narrow gorges and valleys, heavy rainfall and tropical humid climate have made this area with an environment most ideal for luxurient plant growth.

Since majority of the ferns are highly sensitive and recess in undisturbed evergreen forests, even a slight ecological perturbation may affect its very existence. The forests of South India have recently become vulnerable to ecological imbalance due to various anthropogenic activities. Consequent to this detrimental biotic interference on the natural ecosystem, many sensitive species that have existed a decade ago are feared to be devastated from this area. Inspite of several botanical explorations by various groups, many species that were collected in the recent past could not be traced out.

Though several taxonomists like Bir (1962-1991), Nayar and his students (1954-1993) and Manickam. (1976-1992) have contributed much on the understanding on the pteridophytic flora of India, Beddome's (1864-1892) work still remains as the only fully illustrated and described account on the fern flora of India. Beddome's descriptions mostly contain only those aspects of external morphology which were regarded as of taxonomic and diagnostic importance a century ago. Since the taxonomy and nomenclature of pteridophytes were more confused than those of any other group of plants, Beddome's accounts are today of very limited utility only. Beddome's similar work on ferns of Malaya and Ceylon have since been updated by later workers like Holttum and Sledge respectively. However, such an attempt has not been made in the Indian sub-continent. Being a vast country rich in pteridophytes, it will be more practical if detailed studies are undertaken of selected groups or families of particular region, with an objective of preparing a revised pteridophytic flora of India by compiling all such works. It is in this context, we have undertaken the detailed morphological and taxonomical studies on the polypodioid ferns of South India.

The polypodioid ferns constitute a major component of the fern flora of India. Inhabiting the tropical or warm-temperate forests they constitute the largest group of extant ferns. They flourish luxuriantly in the interior of dense humid forests primarily as epiphytes, making access to them formidable. They can be easily distinguished from related groups by predominantly epiphytic, frequently lithophytic habit; having a siphonostelic or dictyostelic vascular cylinder; ground tissue often with scattered sclerenchyma strands; usually reticulate venation; exindusiate sori and sporangia with typical leptosporangiate development. Because of the great diversity of habit and form, this group has received a considerable attention from fern taxonomists, cytologists and morphologists. However, they still remain as one of the taxonomically difficult groups.

Except for a few scattered papers dealing on a few selected genera, the only available work on this group as a whole in India is that of Satija and Bir (1985). However, their work is primarily based on Himalayan ferns and without any description and illustration. In the present treatment, all species treated are provided with detailed morphological description, illustration, distribution and ecology, and nomenclatural notes.

Review of literature

Since the polypodioid ferns form a large and ambiguous group, their systematic treatment at family and generic levels, and evolutionary trends are still obscure and arbitrary. However, during the recent past, this group has received considerable importance and attention from the scientific world.

Several monographs and revisions dealing with selected genera of this group have appeared during the current century. The earliest among them is the monograph of the genus *Pyrrosia* by Giesenhagen (1901). He treated fifty species under *Niphobolus* describing their venation, epidermis and indument. In 1935, Ching studied this genus from the Asian mainland, Japan, and Taiwan province of China, described five new species and made several new combinations under it. Recently, Hovenkamp (1986) published a monograph of *Pyrrosia* and recognized fifty one species. Significant contributions on the reclassification of this genus is available from Nayar and Chandra (1967) and Shing (1983). Hozhizaki (1981) made extensive studies on the cultivated species of *Pyrrosia*.

Copeland (1928) published an account of the genus *Leptochilus*, transferring many species of it to various other genera. In 1956, Sledge presented a critical account on the Sri Lankan species of this genus reducing two species to the status of forma and a third species to synonym.

The revision of Grammitidaceae of New Guinea has been published by Copeland (1952). Since the publication of Copeland's (1951) monograph of *Grammitis* various floristic treatments and regional accounts of this genus have been published (Morton, 1967; Sledge, 1967; Bishop, 1977; Parris, 1975,1983,1991; Parris & Given, 1976). The genus *Ctenopteris* in America is revised by Copeland (1955).

Monographs of a few more genera have also been published, namely, *Adenophorus* by Bishop (1974); *Cochlidium* by Bishop (1978); *Aglaomorpha* and *Drynaria* by Roos (1985); *Goniophlebium* by Rodl-Linder (1990); *Microsorum* by Bosman (1991) and *Platycerium* by Hennipman and Roos (1982). Bosman considered *Microsorum* in a strict sense comprising of twenty species of which two are new descriptions. World revisions of *Lepisorus* and *Loxogramme* have been currently undertaken by Zink (Switzerland) and Price (U.S.A) respectively. Price (1984) made three new combinations under *Loxogramme* and published (1990) four new Asian species under it.

Classification of 'Polypodiaceae' was made by Ching (1940) and Alston (1956). Several papers have been published by Pichi Sermolli (1953-1982) dealing with the nomenclature of families and genera of this group. Significant contributions on nomenclature of many genera are available from Ching (1964), Morton (1970), and Bosman (1986). Besides these, several papers have been published on individual species of many genera.

The first illustrated account on the plant wealth of India is that of van Rheede (1678-1703). He included a number of ferns and fern allies of Malabar Coast in his *Hortus Indicus Malabaricus;* which Linnaeus (1753) has referred to while naming them in his *Species Plantarum.* Four species of Polypodiaceae were illustrated and described by Rheede from this region. Since then several authors (Burmann, 1768; Swartz, 1806; Willdenow, 1810; Roth, 1821; Zenker, 1835; Graham, 1836; Agardh, 1839; Fee, 1845-1852; Miquel, 1850-52) either based South Indian plants for their new descriptions or referred them in their work. References of the ferns of South India can also be found in Hooker's (1837-1854) *Icones Plantarum* and *Species Filicum* (1846-1864).

However, much on the understanding of the pteridophytes of South India is by the contributions of Col. R.H. Beddome (1864-1892). He made extensive collections and based on this observations he published *Ferns of Southern India* (1864) and *Ferns of British India* (1866). Following this, he prepared the *Handbook to the Ferns of British India, Ceylon and the Malay Peninsula* (1883). Beddome (1864) illustrated thirty species of polypodioid ferns from South India. Still now, Beddome's work remained as the most important and useful reference in the taxonomy and identity of Indian fern flora. Since then, the concept of various taxonomic groups of ferns have considerably changed, later authors (Nayar & Kaur, 1974; Chandra & Kaur, 1987) have made an attempt to provide nomenclatural equivalents to the plants described by Beddome. Sledge's (1960) account on the Polypodiaceae and Grammitidaceae of Ceylon is also a source of valuable information on the understanding of this group in South India.

However, in the recent past, there is a great spurt in pteridological research in India, and several floristic accounts have been published. They include *Fern Flora of Meghalaya* by Baishya and Rao (1982), *Fern Flora of Palni Hills* by Manickam (1986), *Pteridophyte Flora of Western Ghats-South India* by Manickam and Irudayaraj (1992), and *Fern Flora of Malabar* by Nayar and Geevarghese (1993).

Revisions of many genera of polypodioid ferns in India or in restricted areas have been attempted by several authors. *Drynaria* and *Pseudodrynaria* in India is revisited by Nayar (1961 a); *Microsorum* by Nayar (1961b), Bir and Trikha (1968a), Madhusoodanan and Nampy (1993); *Pyrrosia* by Nayar and Chandra (1965), Satija *et al.* (1983); *Phymatodes (=Phymatosorus* and *Phymatopteris)* by Bir and Devi (1968); *Pleopeltis* by Bir and Trikha (1968b),*Polypodium lineare* complex and allied species by Bir and Trikha(1969); *Polypodium* and *Goniophlebium* by Bir *et al.* (1974); *Lepisorus excavatus* group by Bir and Trikha (1974); *Lepisorus kashyapii* in the western Himalayas by Bir and Satija (1981).

Contributions on the detailed morphology of many polypodioid ferns have been made by several workers. They include studies on *Drynaria* by Nayar and Kachroo (1953), Nayar (1958); *Pseudodrynaria* by Nayar (1954); *Loxogramme* by Nayar (1955); *Drymoglossum (= Pyrrosia)* by Nayar (1957); *Pyrrosia* by Nayar (1961c); *Microsorum* by Nayar (1963b), Nayar and Madhusoodanan (1977); *Leptochilus* and *Paraleptochilus* by Nayar (1963a). Preliminary observations on the polypodioid ferns of South India was published by Nampy and Madhusoodanan (1992).

Several new taxa belonging to the polypodioid ferns of India have been published, namely, *Phymatodes (= Phymatosorus) banerjiana* by Pal and Pal (1963), *Kaulinia (= Microsorum)* by Nayar (1964), *Grammitis pilifera* by Ravi and Joseph (1979), *Nistarika (= Leptochilus)* by Nayar *et al.* (1985), *Phymatosorus beddomei* by Ghosh (1985), and *Microsorum (= Phymatosorus) malabaricum* by Nayar and Geevarghese (1993). There are also a few records of taxa new to India by Balakrishnan and Chowdhury (1967), Bhargavan (1973), Dixit *et al.* (1976) and Bhargavan and Nair (1981).

The morphology of gametophytes have received considerable attention since long. Stone (1960) made critical observations on the gametophytes of *Grammitis* and *Ctenopteris*. Other references pertaining to the study of gametophytes of polypodioid ferns include Nayar (1961d, 1962, 1965, 1968) and Nayar and Raza (1970). The contributions of Atkinson and Stokey (1964), Nayar and Kaur (1971) and Mehra and Gupta (1986) are a few elaborative work on the gametophyte morphology of homosporous ferns.

Cytological investigations on some species of this group have been made by earlier workers like Fabbri (1957, 1963, 1965), Chairugi (1960), Abraham *et al.* (1962), Abraham (1963), Evans (1963), Ghatak (1963), Lloyd (1963), Bir (1965), Kuriachan (1965, 1968, 1973), Bhavanandan (1968), Bir (1973), Bir and Trikha (1979a), Vasudeva and Bir (1983), Manickam (1984), Manickam and Irudayaraj (1988), and Sankari Ammal and Bhavanandan (1991). Hartman (1931) has studied the mechanism of antheridial dehiscence and Chandra (1962) made observations on the ontogeny of peltate palea in the Polypodiaceae.

Significant contributions to the better understanding of stelar pattern in polypodioid ferns are provided by Chandra (1982) and Molly (1984). Besides these, information on the vascular morphology of this group are available from contributions of Nayar and his students. Venation pattern is highly useful in defining many genera. Mitsuta (1983, 1984 a & b) has made detailed studies on the venation and systematics of Polypodiaceae. Similarly, structure and histochemistry of stomata and epidermal cells in a few species of Polypodiaceae have been attempted by Patel *et al.* (1975).

Since spores possess a unique morphology, they are highly useful in understanding the taxonomy and phylogeny of various groups. The spores of polypodioid ferns are either monolete or trilete with ornamentations on exine. Contributions to the spore morphology of this group have been made by Nayar (1963c), Nayar and Devi (1964,1965), Devi (1977, 1978,1981,1988), Lloyd (1981) and Uffelen (1993).

Area of study

South India, the area of study, includes states of Kerala, Tamil Nadu, Karnataka, Andhra Pradesh and union territories of Mahe and Pondicherry (Fig. 1). Geographically the area falls between 8° and 20° North latitudes and 74° and 85° East longitudes . It covers an area of 4,67,186 sq.km. and is bordered on the north by Maharashtra, Madhya Pradesh and Orissa, on the east by the Bay of Bengal, on the south by Indian Ocean and on the west by Arabian sea.

South India can be divided into two floristic regions, the Malabar and the Deccan. Malabar is a long and narrow strip of land running parallel to the coast of Arabian sea (Malabar Coast), west of the Western Ghats. This region is floristically rich and includes coastal plains and a series of hill ranges of the Western Ghats traversing the states of Gujarat, Maharashtra, Goa, Karnataka, Kerala and Tamil Nadu. The more or less continuous hill ranges of Western Ghats have a major discontinuity in the Palghat gap separating the Nilgiri ranges from the Anamalais. The Ghats descend steeply in the west facing the Arabian sea whereas they merge gradually through a series of hills with the Deccan Plateau in the east. The Western Ghats mountain ranges have an average altitude of 1550 m but rises to 2000-2500m peaks here and there. Anamudi, the highest peak (2695 m) south of the Himalayas, flanks on the Western Ghats. It supports tropical moist evergreen forests and moist deciduous forests on the western slopes. Of the total of about 800 species of Indian ferns, nearly 250 occur in South India and well over 175 of them are found in the middle region of West Coast where the rainfall is high (Nayar and Geevarghese, 1993).

The Deccan is the great tableland of the South Indian Peninsula, stretching from the Aravallis, the Malwa, the Vindhyas, the Satpura and the Chota Nagpur hills in the north and almost right down to Kanyakumari in the south. The Western and Eastern Ghats borders it on either side. The northern part of the Plateau slopes westwards while the southern part slopes towards the south-east. It is the rain shadow region of the Western Ghats and is characterised by the tropical deciduous forests while in the open plains it is replaced by drought resistant species and thorny shrubs. The eastern coastal plains of the Deccan Plateau is a broad strip running parallel to the coast of Bay of Bengal (Coromandel Coast) and gradually rising from it. It consists of fertile coastal plains mainly formed by the deltas of Cauveri in Tamil Nadu, Godavari and Krishna in Andhra Pradesh and a number of small rivulets and streams. The Eastern Ghats, run along the eastern border of the Deccan Plateau, are less prominent and less continuous than the Western Ghats. The continuity of the Eastern Ghats is broken by the Godavari river trough which cuts across them from west to east.

The Malabar and the Deccan regions together provide a wide variety of climatic

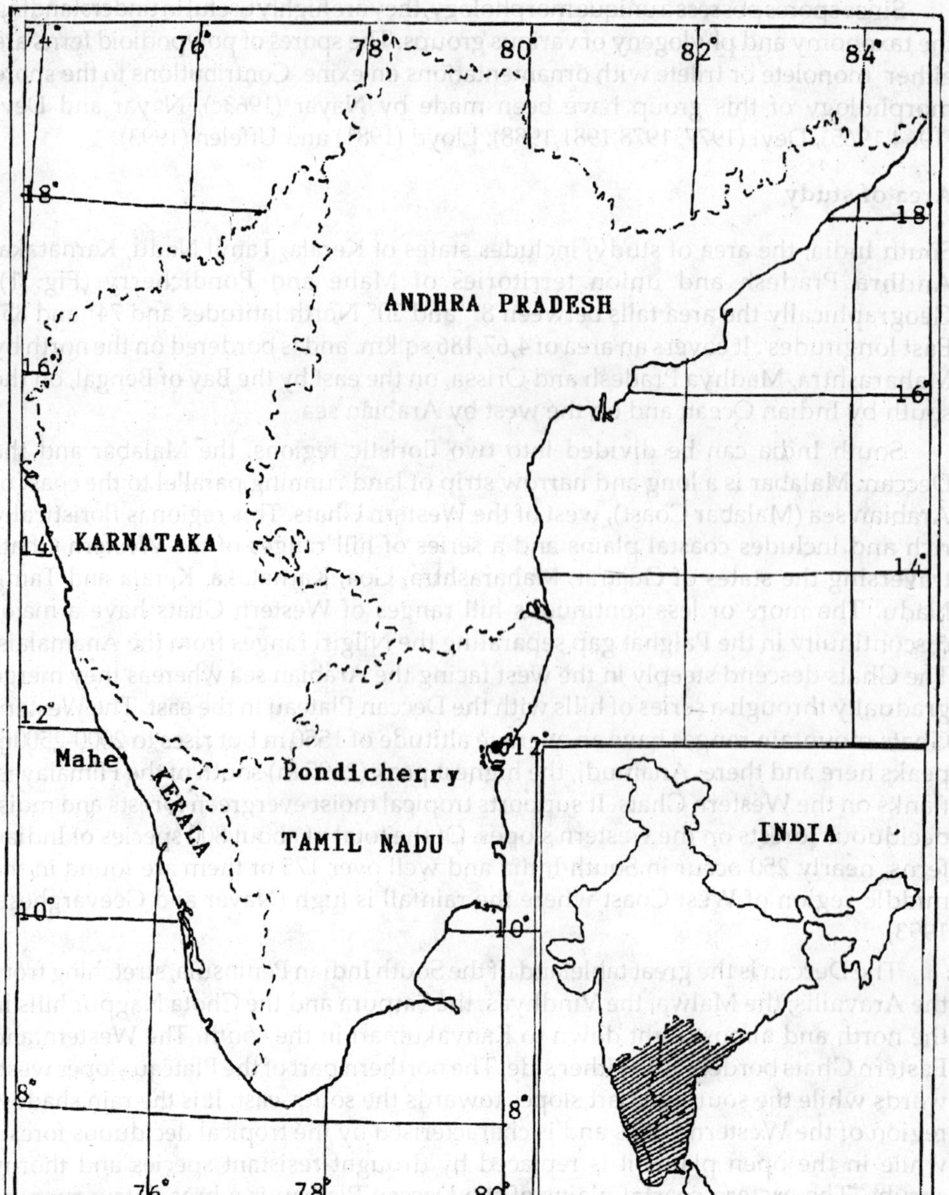

Fig. 1. Map of the study area

and edaphic zones with mountain ranges, hillocks, valleys, swamps, marshy lowlands, sandy sea-coasts, fresh water streams, rivers, ponds and back waters on the sea-front and harbours diverse type of vegetation.

The major South Indian hills having an altitude of more than 1800 m are the Palni (Kodaikanal, 2328 m), Nilgiri (Dodabetta, 2621m; Ootacamund, 2245m), Bababudangiri (1925m),Anamalais (2678 m), and Cardamom hill ranges (Meher-Homji, 1967). On account of the comparatively low temperature prevailing in the South Indian hills, the climate and vegetation of these areas are often spoken of as temperate type. However, Meher-Homji (1965) has described the climate of this region as tropical montane type.

South India receives both north-east monsoon (October-December) and south-west monsoon (June-September). The former is more active in Tamil Nadu, Pondicherry and Andhra Pradesh, while the latter is more vigorous in Kerala and coastal Karnataka. The West Coast of South India receives the heaviest rainfall with more than 2200 mm annually, whereas Andhra Pradesh and Tamil Nadu receive only 1000-2000 mm rain fall, and southern parts of Andhra Pradesh get only 500 mm annually. Rain fall is the chief climatic factor that controls the distribution of ferns in the Western Ghats.

The major rivers of South India are Krishna, Cauvery and Godavari. They originate in the Western Ghats, flows to the East Coast and finally empties into the Bay of Bengal. While small rivers like Netravathi, Bharathapuzha, Pampa and Periyar originate in the Western Ghats, flows westwardly and pour into Arabian sea. These rivers are mainly rainfed and many of them shrink into rivulets during summer.

The climate in South India is in general megathermal (Subramanyam *et. al*,1965; Rao *et al*, 1972). Chowdhury and Sarwade (1982) classified the homoclimatic regions of India into 5 categories, namely, arid region, semi-arid region, sub-humid region, humid regime and super-humid regime. Among them, South India falls under four homoclimatic types. The coastal districts of Andhra Pradesh, interior area of Karnataka and some districts in Tamil Nadu come under semi-arid climate. Northern coastal Andhra Pradesh, southern districts of Karnataka and northern Tamil Nadu experience dry, sub-humid climate while coastal Karnataka and northern Kerala have moist sub-humid type of climate. The humid regime predominates in southern districts of Kerala and at higher elevations around Coonoor and Ooty (Tamil Nadu), whereas super-humid regime occupies only at Kodaikanal (Tamil Nadu).

Distribution and ecology

The polypodioid ferns with about 76 genera and 1500 species are world wide in distribution, but mostly confined to the tropics. In South India, they are represented by 14 genera and 33 species. They are either epiphytic or epilithic and most of them are restricted to undisturbed humid evergreen forests of medium to higher elevations. A few members like *Drynaria quercifolia*, *Pyrrosia heterophylla* and *P. lanceolata* are even adapted to live in dry exposed situations at low altitudes. The dense vegetation between 500-2500 m altitude, is by far, richest in polypodioid ferns.

Since the grammitid ferns are precise in their habitat requirements, they are highly restricted in their distribution. A few like *Prosaptia contigua, P. obliquata,* and *Ctenopteris perplexa* grows along with *Hymenophyllum gardneri, Asplenium zenkeranum, Selaginella involvens,* etc. on well shaded, large boulders in the water courses or on the spray zones of water falls which never dry up during the hot months. They have been collected only from Tirunelveli region in South India. While species of *Grammitis* have been collected from the Nilgiris, Tirunelveli mountains, Agasthiamala and Wynad. They occur on partially shaded or exposed rocks and tree tunks in humid localities at higher elevations.

The family Loxogrammaceae is represented by a single genus *Loxogramme* with two species in South India of which *L. chinensis is* collected only from a few localities like Naduvattam (TN), Banasuramala and Vellarimala (Kerala) while *L. cuspidata* have a wider distribution. They occur on moss covered wet rocks or tree trunks (as low epiphyte) above 900 m altitude.

Microsorum pteropus grows epiphytically on exposed tree roots or on rocks in mountain streams. It can tolerate quite long periods of floods in submerged state. *Leptochilus metallicus* and *L.thwaitesianus* are ferns of this habitat, but grow on well shaded rock crevices beside watercourse. Other species of *Leptochilus* grow on the bases of woody shrubs or trees in the forests, all are fairly common in South India. Ferns of the montane habitat include *Lepisorus amaurolepidus, L. nudus, Microsorum membranceum, Phymatomeris montana, Phymtosorus beddomei, Pleopeltis macrocarpa, Pyrrosia ceylanica,* and *P. porosa*. They inhabit on moss covered rocks or tree trunks in dense evergreen forests. *Pyrrosia heterophylla* and *P. lanceolata* also occur on trees in the coastal areas having a salt-laiden air.

Five species are endemic to South India, namely, *Grammitis austroindica, G. pilifera, Leptochilus bahupunctika, Phymatosorus beddomei* and P. *malabaricus,* of which *L. bahupunctika and P. malabaricus* are reported from Kerala only. Similarly, *Grammitis, Microsorum* and *Nistarika (=Leptochilus)* are facing serious threat of extinction (Madhusoodanan, 1991).

Phlebodium aureum and *Phymatosorus scolopendria* do not occur wild in South India but widely grown as an ornamental. Escapes from cultivation of these species are found growing wildly near human habituations.

Since most species require specific environmental requirements and show narrow range of adaptability and tolerance, only a few species survive in *ex situ* cultivation even in botanical gardens.

Materials and methods

Specimens for the present study were collected from different parts of South India and for most of the species, characters were studied by observing live plants in the field . Extensive field studies have been conducted in different parts of South India to understand the ecological niches, infra specific variation, biological associations and distributional pattern of each taxa. Several species collected from wild are cultivated in the botanical garden of Calicut University. Materials for herbaria were pressed using conventional methods in the collection site itself. Herbaria prepared during the present work were deposited at Calicut University Herbarium (CALI).

Specimens kept at CALI and Madras Herbarium (MH), Coimbatore and type photographs or xerox copies of type specimens received from the following herbaria were also consulted (abbreviations follow the Index Herbariorum).

BM	–	British Museum (Natural History), London
K	–	The Herbarium and Library, Royal Botanic Gardens, Kew, U.K.
L	–	Rijksherbarium, Leiden, The Netherlands.
MA	–	Real Jardin Botanico, Madrid
MICH	–	University Herbarium, University of Michigan, Ann Arbor, U.S.A.
P	–	Museum National d Histoire Naturelle, Laboratoire de Phanerogamie, Paris.

Comparative morphological study was done using all available specimens of a species. While representative sampling of specimens were made for studying palea, anatomy, venation, sporangia, paraphyses and spores. Anatomical studies of most species were done using fresh materials. In the case of dried materials, they were softened by boiling in dilute (50%) glycerine. Transections of rhizome were made at the internodes and of stipes from 1 cm above the phyllopodium. The sections were made semi-permanent by mounting in glycerine.

Palea were taken from the extreme apex of the rhizome or from its lateral buds (for young) and several centimeters behind the apex of the rhizome (for mature) Much care was taken to get the palea intact to decide whether it is basifixed, pseudo-peltate or peltate. Average size is noted by measuring different samples. Observations were made using Leitz stereo microscope. The venation pattern of most of the species was studied by soaking the lamina with 95% alcohol, and observed this in transmittent light. Coriaceous leaves were cleared by keeping the pinnae in 5-10% of KOH with a few drops of Basic Fuchsin at 60°C overnight. After clearing, they were washed several times and mounted in glycerine. Camera lucida drawings were made using Leitz stereo microscope.

Sori of different developmental stages were observed for sporangia and paraphyses. They were separated from each other and mounted in glycerine. The shape and position of receptacle was noted after removing the sori. Palynological observations were made after sieving the spores through a wire mesh. They were then soaked either in 95% alcohol or acetolysed (Erdtman, 1952,1957) and mounted in glycerine jelly. Using Carlzeiss Jena microscope, a representative selection of the spores of each species were studied. Gametophytes of selected species were studied by spore culturing method (Nayar, 1962). Fresh spores were sown on sterile agar medium supplemented with modified Knop's solution. The culture was kept in culture room under controlled conditions and various developmental stages were drawn. Gametophyte morphology, chromosome numbers and etymology for most of the species were adopted from available literature.

For each species, there is an illustration of a whole plant (habit) together with details of venation, palea, indumentum, sporangium, paraphysis and spore. Except the habit sketch, all others were drawn using camera lucida. Descriptions and illustrations were done using fresh materials except in a few cases. The ecological notes are based on our field notes and informations collected from herbarium sheets.

The distribution of each species in South India is mapped. Native names are given when known. Nomenclatural notes are provided on infraspecific variations and relationships.

Plan of the work

The introductory part begins with the importance of the present work, followed by review of literature, salient features of the area of study, distribution and ecology and materials and methods.

In the systematic part, classification of the polypodioid ferns and an artificial key to families are provided. The families are treated in the alphabetical order. Each family is provided with a historical review followed by its description. Key to genera are presented under each family and the following informations are given for each genus: a brief general account comprising its circumscription and etymological derivations of the name, relevant bibliographical data, type species, synonymy, brief description, gametophyte morphology, chromosome numbers, distribution and ecology followed by key to the species. The genera under each family and species under each genus are treated in the alphabetical order. For each species the following informations are provided: correct nomenclature, relevant synonyms, detailed morphological description, distribution and ecology and nomenclatural notes. List of specimens examined from the area of present study are given at the end of each species and are cited by state, district, and place in the alphabetical order. Distribution maps and illustrations are provided for most of the species and photocopies of type specimens are provided wherever necessary. A discussion on comparative morphology and its bearing on phylogeny of polypodioid ferns is given. The bibliography on literature cited and an index to scientific names are provided.

SYSTEMATIC TREATMENT

Polypodioid Ferns

The concept of polypodioid ferns is highly argumental. At one time they were grouped in one of the largest families of plant kingdom, the 'Polypodiaceae' (Fee, 1850-1852; Hooker & Baker, 1865-1868; Diels, 1899; Christensen, 1906) which include almost all ferns characterised by "vertical incomplete annulus of sporangium opening by a transverse slit". As a result, it included over 200 genera and 7500 species, most of them pertaining to the present Filicopsida. It was Ching (1940) who first attempted a comprehensive taxonomic treatment of the family. He was aware of the arbitrariness of the group and rightly stated "Polypodiaceae, as accepted now, seems to be a bottomless pit into which any fern species or genus which proves, in point of sporangia, incompatible with other families can be thrown, regardless of the differences in other respects...". Consequently, he split the 'Polypodiaceae' into 33 families and paved a way for modern classification of the fern families. However, Copeland (1941) strongly objected Ching's (*l.c.*) view particularly for the "increase in the number of families, beyond the possibility of clear definition, entails unavoidable increase of inconvenience and error". He expressed his opinion while stating "we can comply perfectly, with 'the principles of modern plant taxonomy' by combining families, instead of splitting and chopping them up...".

Ching (1940) placed the polypodioid ferns under six families, namely, Loxogrammaceae, Cheiropleuriaceae, Dipteridaceae, Platyceriaceae, Polypodiaceae and Grammitidaceae. Since then, there has been considerable difference of opinion among various authors, on the family or generic concept of polypodioid ferns. Copeland (1947) placed all these ferns in a single large family Polypodiaceae; Holttum (1947, 1949) in two families Polypodiaceae and Grammitidaceae; Wagner (1961) in three families, *viz.*, Dipteridaceae, Cheiropleuriaceae and Polypodiaceae; Pichi Sermolli (1958, 1959) and de la Sota (1973) in four families : Dipteridaceae, Cheiropleuriaceae, Polypodiaceae and Grammitidaceae; Wilson (1958), Nayar (1970), and Pichi Sermolli (1977) in five families : Dipteridaceae, Cheiropleuriaceae, Polypodiaceae, Loxogrammaceae and Grammitidaceae. Wagner's (*l.c.*) Polypodiaceae consisted of three subfamilies, namely, Polypodioideae, Grammitidoideae and Loxogrammoideae. From these systems, the most natural classification seems to be given by Pichi Sermolli (1977). He proposed a common origin for the whole group and assembled all polypodioid ferns in a single order Polypodiales. It is placed in between Gleicheniales and Matoniales and distinguished two suborders : Dipteridineae and Polypodineae. The former appears to be more primitive and includes two families, Dipteridaceae and Cheiropleuriaceae. The Polypodineae have three families : Grammitidaceae, Loxogrammaceae and Polypodiaceae. In South India 14 genera and 33 species of polypodioid ferns occur under three families (Grammitidaceae, Loxogrammaceae and Polypodiaceae).

Key to the families

1a. Stipe articulate ; spores monolete.............................*Polypodiaceae*

1b. Stipe usually non-articulate ; spores mostly trilete..........................2

2a. Veins anastomosing; sporangial stalk 2-rowed for most of its length.............*Loxogrammaceae*

2b. Veins not anastomosing; sporangial stalk 1-rowed for most of its length..*Grammitidaceae*

GRAMMITIDACEAE Newman

The family Grammitidaceae was first recognised by Newman (1840). However, Newman's Grammitidaceae largely remained unnoticed until recently. Meanwhile, Ching (1940), apparently unaware of Newman's Grammitidaceae raised the tribe Grammiteae of Presl (1836) to the rank of a family and used the name Grammitaceae. Many taxonomists like Pichi Sermolli (1974), Satija and Bir (1985) followed Ching (*l.c.*), while several others (Holttum, 1954; Parris, 1983) considered Ching's name of Grammitaceae as an error and used the name Grammitidaceae to agree with the type genus *Grammitis* Swartz. Since then, there exists great controversy regarding the acceptance of the name Grammitaceae or Grammitidaceae. Proposal has been made by Panigrahi (*Taxon* 35 : 604. 1986) to conserve the family name Grammitaceae against Grammitidaceae. The Committee for Pteridophyta has examined his proposal and concluded that the correct citation for the name is Grammitidaceae Newman (*Hist. Brit. Ferns* 1 : 7. 1840) which antedates Grammitaceae (C. Presl) Ching by about a century. The Committee has not recommended Panigrahi's proposal and expressed for the preference for original spelling Grammitidaceae over Grammitaceae (*Taxon* 42 (1) : 120. 1993).

The modern regional fern floras of the Old World treat Grammitidaceae as a separate family (Holttum, 1954; Copeland, 1960; Schelpe, 1970; Devol, 1975; Manickam & Irudayaraj, 1992). However, in the recent New World floras, it is kept as a tribe (Grammiteae) of Polypodiaceae (Proctor, 1977; Tryon & Tryon, 1982). In the present treatment, it is kept as a distinct family. The Grammitidaceae is characterised by fronds generally not jointed to the rhizome, veins usually free and without included veins, sporangia with one-rowed stalk for most of its length and trilete, chlorophyllous spores.

Ching (1940) recognised two tribes under Grammitidaceae, *viz.*, tribe Grammiteae and tribe Cochlideae. However, subsequent authors (Nayar, 1970; Pichi Sermolli, 1977; Parris, 1991) have not followed Ching in maintaining these tribes. Grammitidaceae as construed by Ching (*l.c.*) include nine genera, namely, *Grammitis* Swartz (with which *Ctenopteris* Bl. ex Kunze was synonymised), *Calymmodon* Presl, *Acrosorus* Copeland, *Prosaptia* Presl, *Holcosorus* Moore, *Cochlidium* Kaulf., *Scleroglossum* v.A.v.R., *Nematopteris* v.A.v.R., and *Oreogrammitis* Copeland. Excluding *Holcosorus* (now this has been transferred to Polypodiaceae, Pichi Sermolli, 1977), all other genera are still retained in Grammitidaceae along with some additional genera, *viz.*, *Glyphotaenium* J. Sm., *Adenophorus* Gaud., *Ctenopteris* and *Xiphopteris* Kaulf. The above twelve genera have been accepted in most of the Old World fern floras while in the New World floristic accounts only a few of the above genera has been recognised.

Grammitidaceae Newman, Hist. Brit. Ferns 1 : 7. 1840.

Type genus : *Grammitis* Swartz

Grammiteae Presl, Tent. Pterid. 205. 1836 as tribe.

Grammitaceae (Presl) Ching, Sunyatsenia 5 : 264. 1940.

Rhizome erect or creeping, usually paleaceous, simple or dissected solenostelic. Palea shining, entire or ciliate. Fronds tufted, usually non-articulated, with or without acicular hairs, simple or pinnatifid or pinnate. Veins free, without included veins. Sori exindusiate, receptacle round or elongate, superficial or submarginal, raised or sunken in the leaf tissue, receptacular bristles sometimes present. Sporangia setaceous or not so; sporangial stalk usually one-rowed for most of its length; annulus 11-15-celled. Spores trilete, tetrahedral or globose, chlorophyllous, granulose to somewhat verrucate.

Notes : In South India three genera and eight species are found, of which two species (*Grammitis austroindica and G. pilifera*) are endemic. The genera with their respective number of species are *Ctenopteris* (2), *Grammitis* (3) and *Prosaptia* (3). The South Indian members of this family occur in mountain forests above 900 m altitude, usually epiphytic or occasionally as epilithic. Because of the extreme habitat requirements, only a few species succeed under cultivation.

Key to the genera

1a. Fronds simple........................*Grammitis*

1b. Fronds pinnatifid.....................2

2a. Sori immersed in bilipped cavities; margins of the palea usually ciliate............................*Prosaptia*

2b. Sori superficial ; margins of the palea usually entire.........................*Ctenopteris*

CTENOPTERIS Blume

It was Blume (1828), who first used the name 'Ctenopteris' for a polypodiaceous genus. He listed several species of *Polypodium* under it without making new combinations. Blume's *Ctenopteris* largely remained unnoticed until Kunze (1846–1848) recognised it and formally made new combinations of the names in his *Bot. Zeit.* However, he ascribed the genus to Blume (1828) and cited *Ctenopteris venulosa* as its type.

Eversince its establishment, the generic identity of *Ctenopteris* is a matter of great discord. Many authors like Copeland (1947, 1952, 1955), Holttum (1954), de la Sota (1973), Nayar and Kaur (1974), Devol (1975), Parris (1983), Chandra and Kaur (1987) and Manickam and Irudayaraj (1992) considered it as a distinct genus. While many others considered it as congeneric with *Polypodium* L. (Beddome, 1864, 1866, 1883; Christensen, 1906, 1938) or under *Grammitis* Swartz (Ching, 1940; Tryon & Tryon, 1982; Parris, 1991). However, *Ctenopteris* is a clearly definable genus and is characterised by usually non-articulate stipe, hairy or glabrate fronds contracted at base, simple veins, mostly 1-rowed sporangial stalk and globose-tetrahedral spores. *Polypodium* differs from it in having articulate stipe, paleate or naked fronds usually broad at base, forked veins, mostly 3-rowed sporangial stalk and bilateral spores. Similarly, *Ctenopteris* can be segregated from *Grammitis* in having non-setose sporangia and pectinate fronds as against usually setose sporangia and simple, entire fronds in *Grammitis.* The genus is treated here as conceived by Copeland (1947).

The name 'Ctenopteris' is derived from two words 'ktenos', a comb and 'pteris', fern; the whole thus referring to the comb-like arrangement of pinnules on the rachis.

Ctenopteris Blume, Fl. Jav. 132. 1828; Kunze, Bot. Zeit. 425. 1846; Copeland, Gen. Fil. 218. 1947; Holttum, Fern Fl. Malaya 2 : 224. 1954, in part; Devol in Li *et al.* (eds.), Fl. Taiwan 1 : 217. 1975, in part.

Type species : *Ctenopteris venulosa* (Bl.) Kunze

Rhizome short, erect, densely paleaceous. Palea ovate-lanceolate, margins entire or with a few hairs at the apex. Fronds tufted, pinnatifid, lanceolate; stipe hairy, non-articulated; pinnae broadly adnate to the rachis, herbaceous, hairy abaxially; rachis hairy on both surfaces, hairs reddish. Venation simply-pinnate, veins free or occasionally reticulate. Sori superficial, terminal on the veins, round or elliptic; receptacle elongate; without paraphysis. Annulus 11 or 12-celled, sporangial stalk 1-rowed. Spores trilete, globose-tetrahedral, granulose.

Gametophyte : The gametophyte of *Ctenopteris* have a prolonged filamentous stage which become ribbon-like when mature (Stone, 1960).

Chromosome numbers : n = 37, 38 (Manton, 1954; Manton & Sledge, 1954; Abraham *et al.*, 1962).

Distribution and ecology : *Ctenopteris* with about 125 species is distributed throughout the tropics ranging from south to New Zealand and north to Tropic of Cancer. Most of its species grow on mossy tree trunks or on wet rocks in mountain forests. Two species, namely, *C. blechnoides* and *C. perplexa* occurs in South India.

Key to the species

1a. Frond hairs all less than 0.5 mm long;
sporangia not mixed with acicular bristles........................*C. perplexa*

1b. Frond hairs all 0.5-1 mm long;
sporangia intermixed with acicular bristles
at least in young stage ....*C. blechnoides*

Ctenopteris blechnoides (Grev.) Wagner & Grether, Univ. Calif. Publ. Bot. 23 : 61.1948; Parris, Kew Bull. 41(3) : 494. 1986.

Grammitis blechnoides Grev., Ann. Mag. Nat. Hist. 2, 1 : 328. t. 17. 1848.

Type : Raiatea, *Sibbald s. n.* (holotype, E?)

Polypodium decorum Brack., U. S. Expl. Exped. 16 : 7. t. 2. f. 2. 1854; Beddome, Handb. Ferns Brit. India 310. f. 165. 1883.

Polypodium blechnoides (Grev.) Hook., Sp. Fil. 4 : 180. 1862.

Cryptosorus blechnoides (Grev.) J. Sm., Hist. Fil. 87. 1875.

Polypodium moultonii Copeland, Philipp. Jour. Sci. 10 : 149.1915.

Ctenopteris moultonii (Copel.) C. Chr. & Tard., Nat. Syst. (Paris) 8 : 181. 1939; Holttum, Fern F1. Malaya 2 : 229. f. 128. 1954; Sledge, Bull. Brit. Mus. (Nat. Hist.) Bot. 2 (5) : 155. 1960.

Rhizome erect to short creeping, densely paleaceous. Palea ovate-lanceolate, 2 × 0.6 mm, medium brown, glabrous but with a few glandular hairs on the apex and margins. Fronds deeply-nearly pinnatifid, 7-16 × 0.3-1.5 cm, lanceolate; stipe non-articulated, 1-16 mm, with occasional simple, 0.2-0.4 mm long, medium red-brown hairs and occasional to scattered glandular hairs with 1-5 branches; pinnules narrowly oblong, entire, basal ones broadly triangular, subcoriaceous, hairs like those on the stipe occasional to scattered abaxially on the costa, sparsely so on both sides of the lamina, margin and costa adaxially. Sori slightly sunken, rim not raised, circular, 3-14 per pinnule on either side of the mainvein. Sporangia intermixed with acicular bristles.

Distribution and ecology : *C. blechnoides* is very rare in South India, being only once collected by Beddome from Thirunelveli hills in Tamil Nadu (Fig. 3). It is also distributed in Sri Lanka, Thailand, Cambodia, Singapore, Peninsular Malaysia, Sumatra, Borneo, Philippines, Sulawesi, New Guinea, Moluccas, Melanesia and Polynesia. It grows either as an epiphyte or lithophyte in shaded localities.

Specimen examined : TAMIL NADU : Kattabomman Dt., Thirunelveli hills, *Beddome s. n.* (K).

Ctenopteris perplexa Parris inedit.

Polypodium parvulum sensu Beddome, Ferns South India 55.t.166. 1864, non Bory, 1810.

Ctenopteris subfalcata sensu Sledge, Bull. Brit. Mus. (Nat. Hist.) Bot. 2 (5) : 156. 1960;

Nayar & Kaur, Comp. Bedd. Handb. Ferns Brit. India 77. 1974; Chandra & Kaur, Nomen. Guide Bedd. Ferns South India & Ferns Brit. India 19, 57. 1987; Manickam & Irudayaraj, Pterid. Fl. W. Ghats 338. pl. 256. 1992. (Fig. 2)

Rhizome short, erect, upto 5 mm in diameter, densely paleaceous at the apex. Palea ovate-lanceolate, 3 × 0.5 mm, uniformly light brown, apex ending in one or two long hairs, margins entire. Fronds tufted, 6.8 × 2.6 cm, narrowed gradually to the base and apex, pinnatifid; stipe non-articulate, 6 × 0.5 mm, hairy on the adaxial surface; pinnules 6 × 2 mm, attached to the rachis by broad contiguous bases, alternate or sub-opposite, gradually tapering towards the tip, apex rounded, margins entire, hairy abaxially, herbaceous; rachis raised on both surfaces, hairy; hairs 0.5-0.9 mm long, unicellular, reddish-brown. Veins simply pinnate, free, not reaching the margin. Sori superficial, 1 mm across, in one row on either side of the midrib, distinct, not sunk in cavities, seated at the end of veins, receptacle elongate, round or elliptic, confluent at maturity; paraphysis absent. Capsule round, annulus 11 or 12—celled, stalk 1-rowed. Spores 27 × 28 µm, globose-tetrahedral, chlorophyllous, granulose.

Distribution and ecology : It is distributed in South India (Fig. 3) and Sri Lanka. *C. perplexa* grows on wet, moss covered rocks or on shaded trees in mountain forests.

Notes : Parris (inedit.) based Sri Lankan specimens for her new species *C. perplexa*. The South Indian materials which were formerly treated under *C. subfalcata* by Sledge (1960) and Manickam and Irudayaraj (1992) or *Polypodium parvulum* by Beddome (1883) also belongs here.

Specimens examined : TAMIL NADU : Coimbatore Dt., Anamalais, *Beddome s.n.* (MH). Kattabomman Dt., Kothayar, *Santhosh Nampy* 45121 (CALI).

GRAMMITIS Swartz

The name 'Grammitis' had its origin from the Greek, 'gramme', meaning a line, possibly referring to their sporangia scattered in a straight line. It was Swartz (1801) who established the genus *Grammitis* but treated under the family Polypodiaceae. He distinguished it from the related *Polypodium* by "capsulae in lineolis rectis, sparsis. Indusia nulla", and included six species under it.

The circumscription and delimitation of *Grammitis* has been a matter of debate since long. Several of the species included by Swartz has been segregated to other genera such as *Xiphopteris, Cochlidium* and *Loxogramme* in the subsequent years. However, two of them, namely, *Grammitis linearis* Swartz (=*G. graminea (Sw.)* Copeland) and *G. marginella* (Sw.) Swartz are still retained in *Grammitis*. Willdenow (1810) and Brown (1810), contemporaries of Swartz, accepted the generic status of *Grammitis*. However, most subsequent authors (Hooker, 1846-1864; Hooker & Baker, 1865-1868) preferred to merge it with *Polypodium*. In 1875, Smith reinstated the genus *Grammitis* to include twelve species, but retained only one of Swartz's original species. But this was not accepted by Christensen (1906-1934). Ching (1940) recognised 300 species under *Grammitis* and placed another genus *Ctenopteris* under it. Since then, Copeland (1947) has studied this complex in some detail and recognized *Xiphopteris, Grammitis* and *Ctenopteris* as distinct genera.

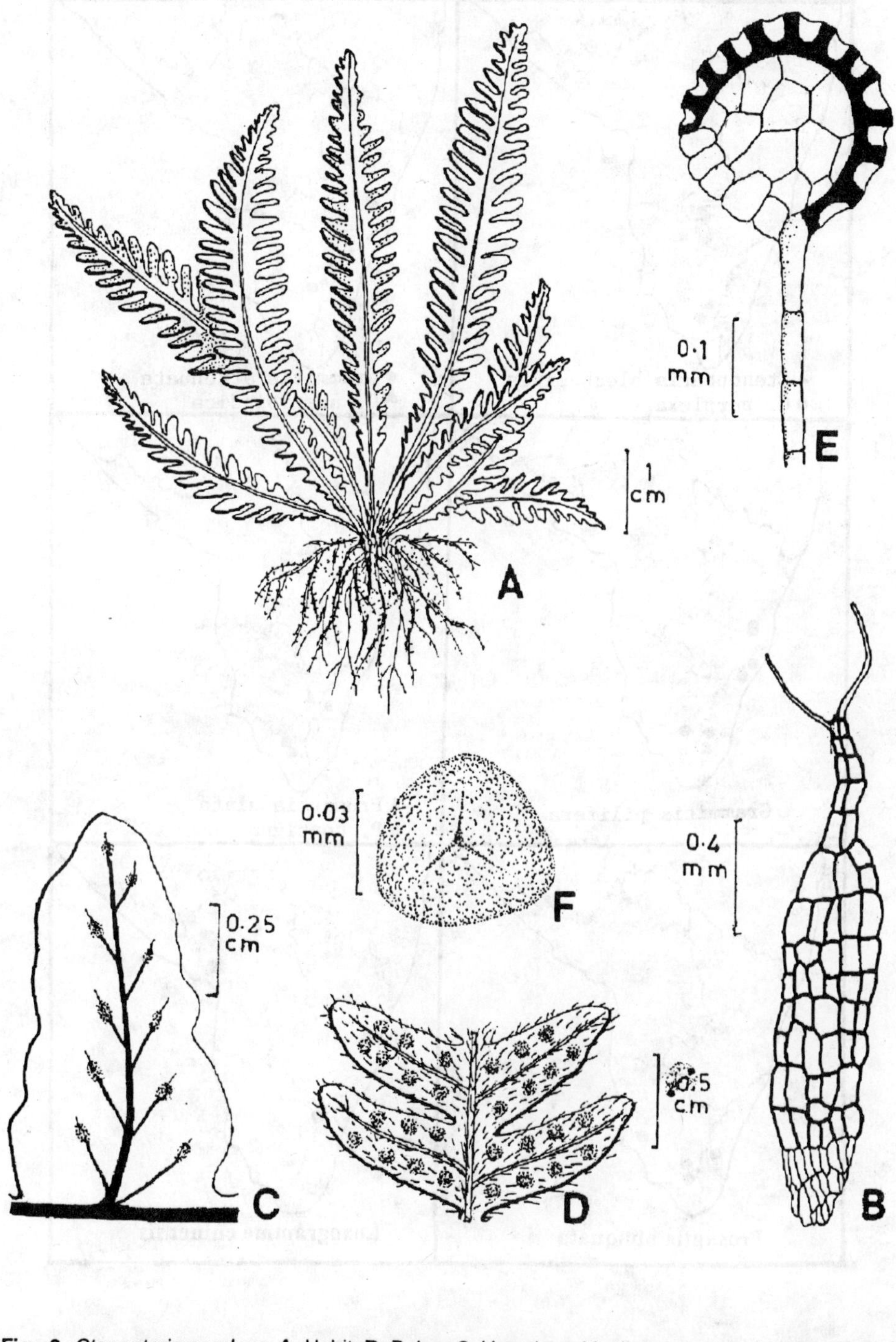

Fig. 2. *Ctenopteris perplexa.* **A.** Habit; **B.** Palea; **C.** Venation of fertile portion of a lamina; **D.** Part of fertile frond enlarged; **E.** Sporangium ; **F.** Spore. (from *Santhosh Nampy* CU 45121)

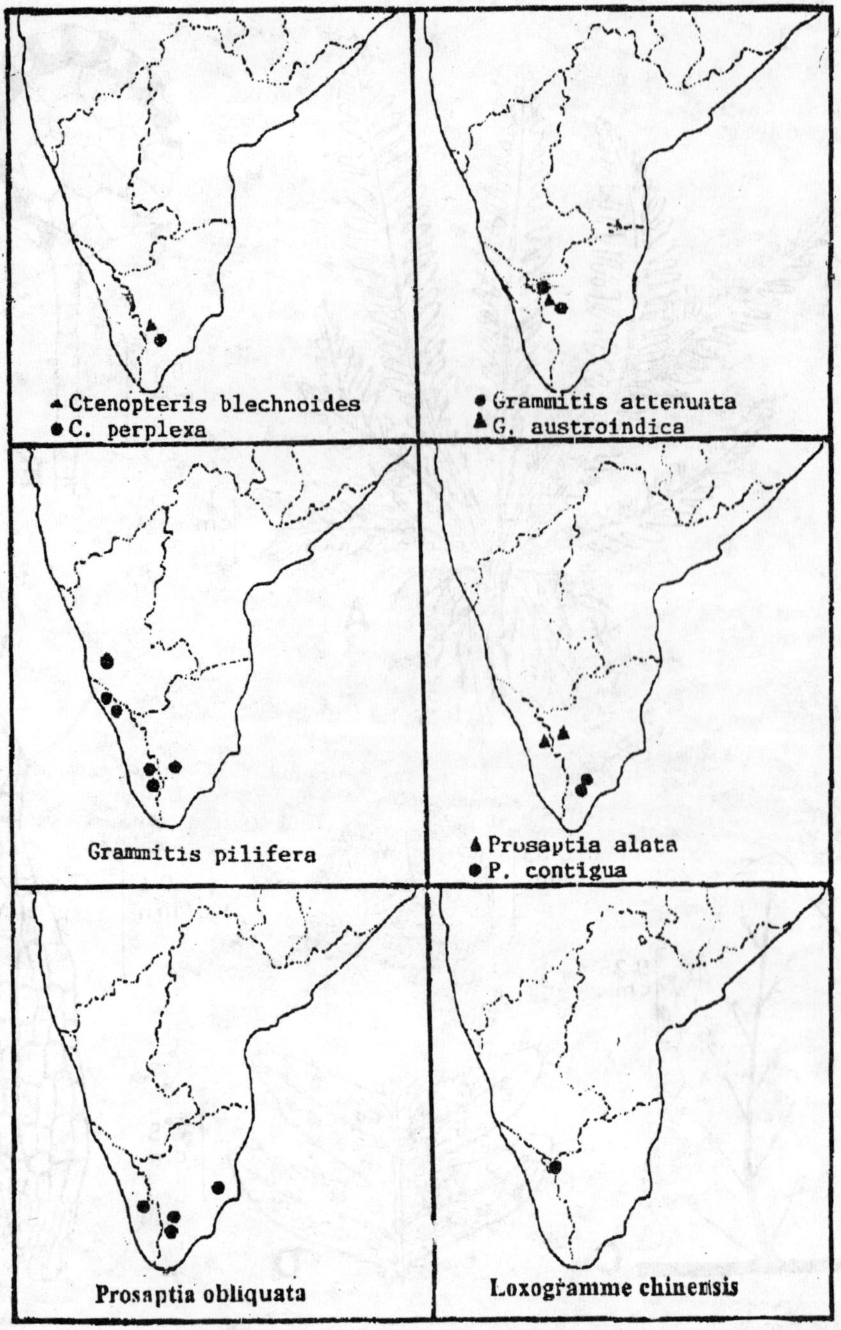

Fig. 3. Distribution maps of species of *Ctenopteris Grammitis, Prosaptia* and *Loxogramme chinensis* in South India

Copeland (1951) maintained the same (*l.c.*) generic delimitation in his monograph of *Grammitis.* He clearly defined the genus as "...characterised by simple, narrow fronds not-articulate to the rhizome, typically entire but sometimes sinuate or shallowly crenate or lobed; normally free veins; exindusiate sori which are round or elongate, normally on the acroscopic branches of veins divergent from the costa; pedicel of the sporangium mostly of a single row of cells; tetrahedral spores containing chlorophyll; the rhizome paleate but the lamina not so". He considered this genus is of Antartic origin and represent a natural group. Parris (1983) in her revision of this genus in New Guinea followed Copeland's (1947, 1951) concept which is accepted in the present treatment.

Grammitis Swartz, Schrader's Jour. Bot. 1800 (2) : 17. 1801; Copeland, Gen. Fil. 210. 1947 & Philipp. Jour. Sci. 80(2) : 93. 1951; Holttum, Fern Fl. Malaya 2 : 213. 1954; Tryon & Tryon, Ferns Allied Plants 747. 1982, in part; Parris, Blumea 29(1) : 13. 1983; Parris in Kubitzki (ed.), Fam. Gen. Vas. Plants 1 : 155. 1991, in part.

Type species : *Grammitis marginella* Swartz, lectotype designated by Christensen, 1906.

Grammitis sect. *Chilopteris* Presl, Tent. Pterid. 208. 1836.

Mecosorus Klotzsch, Linnaea 20 : 404. 1847, in part.

Mecosorus sect. *Chilopteris* Klotzsch, Linnaea 20 : 405. 1847.

Austrogramme Fourn., Ann. Sci. Nat. Bot. Ser. 5(18) : 278. 1873, in part.

Polypodium sect. *Grammiastrum* Fourn., Ann. Sci. Nat. Bot. Ser. 5(18) : 282. 1873.

Oreogrammitis Copeland, Philipp. Jour. Sci. 12 : 64. 1917.

Grammitis subgen. *Melanoloma* Copeland, Philipp. Jour. Sci. 80(2) : 253. 1951.

Grammitis sect. *Grammiastrum* Morton, Contr. U.S. Nat. Herb. 38 : 89. 1967.

Rhizome erect to short creeping or ascending, densely paleaceous all over; solenostelic, stele sheathed by thick-walled dark brown cells. Palea entire, occasionally possess marginal hairs, light yellow to dark brown, glandular basally and apically. Fronds simple, tufted, in two close dorsal rows on the rhizome; stipe short or long, usually velutinous, non-articulated; lamina oblanceolate to lanceolate or linear, entire or undulate or very occasionally lobed, midrib and margins usually hairy, membranaceous to coriaceous. Hairs usually acicular, reddish-brown. Veins free, lateral veins simple or once forked. Sori usually in a single row on either side of the midrib, elliptic or nearly round, usually on the surface of the lamina. Sporangia glabrous or setaceous, intermingled with or without long reddish-brown bristles; sporangial stalk 1-rowed for most of its length. Spores trilete, globose-tetrahedral, chlorophyllous.

Gametophyte : The gametophyte has a prolonged filamentous stage, which later become strap-like, elongated and branched (Stone, 1960).

Chromosome numbers : n = 36, 37 (cf. Parris, 1983).

Distribution and ecology : A genus of about 160 species, *Grammitis* is distributed in South India, Sri Lanka, Malaysia, Taiwan, Japan, Vietnam, China, Australia,

Micronesia, Melanesia, Polynesia, Central and South America, South Atlantic and South Indian Ocean Islands, West Indies, Azores, tropical and southern Africa and Madagascar (Parris, 1983). This genus is represented in South India by three species, namely, *G. attenuata, G. austroindica* and *G. pilifera* of which the last two species are endemic to South India. *Grammitis* grows on mossy tree trunks or on wet, shaded rocks in mountain forests.

Key to the species

1a. Rhizome long creeping .... *G. austroindica*

1b. Rhizome erect to short creeping 2

2a. Palea uniformly dark brown; sporangia generally non-setose and intermingled with long bristles............... *G. attenuata*

2b. Palea uniformly pale brown except for a few central dark cells; sporangia generally setose and without intermingled bristles............... *G. pilifera*

Grammitis attenuata Kunze, Linnaea 24 : 251. 1851; Copeland, Philipp. Jour. Sci. 80(2) : 238. 1951, in part; Sledge, Bull. Brit. Mus. (Nat. Hist.) Bot. 2(5) : 147. t. 3. 1960; Nayar & Kaur, Comp. Bedd. Handb. Ferns Brit. India 75. 1974; Manickam, Fern Fl. Palni Hills 158. 1986; Chandra & Kaur, Nomen. Guide Bedd. Ferns South India & Ferns Brit. India 19. 1987, in part; Manickam & Irudayaraj, Pterid. Fl. W. Ghats 336. pl. 254. 1992; Nayar & Geevarghese, Fern Fl. Malabar 348. 1993, in part.

Type : South India (Nilgiris), Rev. *Schmid s.n.* (holotype, LEIPZIG).

Polypodium parasiticum Mett., Abhhandl. Senckenb. Natur-forsch. Ges. 2 (1) : 36. 1856; Beddome, Ferns South India 55. t. 165. 1864 & Handb. Ferns Brit. India 302. t. 157. 1883, in part.

Grammitis beddomeana Copeland, Philipp. Jour. Sci. 80(2) : 238. 1951, in part, non *Polypodium beddomeanum* Alderw., 1918. (Fig. 4)

Rhizome short creeping or erect, 1-2.5 mm in diameter, densely paleaceous at the apex and less so in older regions; solenostelic, stele sheathed by broad band of thick walled dark-brown cells. Palea oblong or lanceolate, 1-2 × 0.6-0.9 mm, peltate with sub-basal attachment, uniformly dark-brown except a marginal hyaline region, inflated, apex rounded, glandular basally and apically, margins entire. Fronds simple, tufted, 53×5 mm; stipe 7×0.6 mm, terete, non-articulated, velutinous; lamina sword-like or lanceolate, apex blunt or acute, base gradually tapering or acute, subcoriaceous, margins entire or slightly undulate, revolute, scattered-hairy adaxially, margins and midrib frequently hairy, glabrous or occasionally hairy abaxially. Hairs reddish-brown, acicular; stipe hairs short (0.25 mm long), laminar hairs long (above 1 mm). Venation distinct, veins simple or once forked. Sori median, one row on either side of the midrib, either on the acroscopic veinlet of the forked veins or on

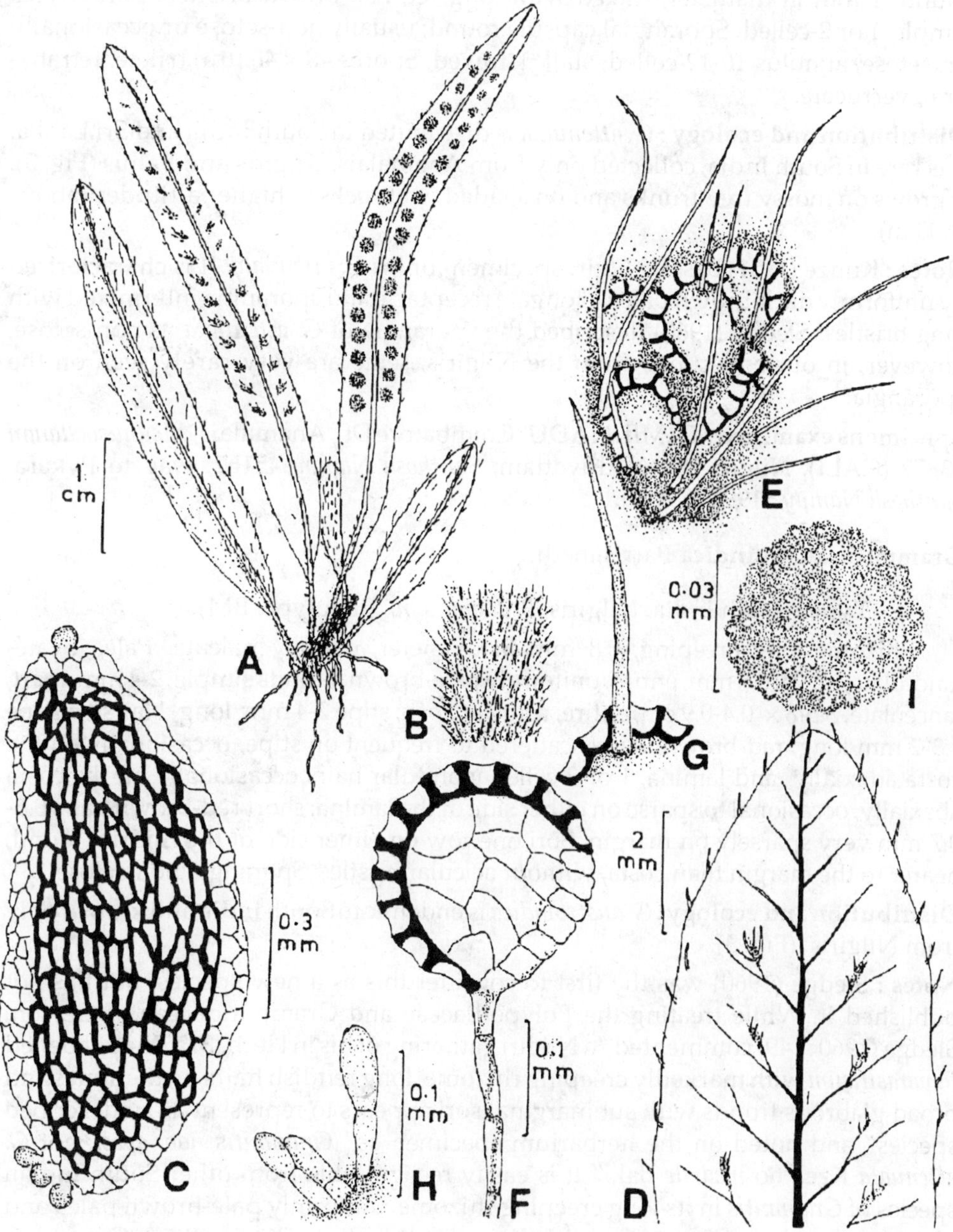

Fig. 4. *Grammitis attenuata.* **A.** Habit; **B.** Portion of stipe enlarged; **C.** Palea; **D.** Venation of fertile portion of a lamina; **E.** Part of Sorus enlarged; **F.** Sporangium; **G.** Portion of sporangium showing seta; **H.** Paraphyses; **I.** Spore. (from *Santhosh Nampy* CU 45162)

simple veins, on the anterior 2/3rd of the lamina, receptacle elongate, mature sorus round, 1 mm in diameter, mixed with long reddish-brown bristles; paraphyses simple, 1 or 2-celled. Sporangial capsule round; usually non-setose or occasionally unisetose, annulus 10-12-celled, stalk 1-rowed. Spores 48 × 48 μm, trilete, tetrahedral, verrucate.

Distribution and ecology : *G. attenuata* is distributed in South India and Sri Lanka. It is rare in South India, collected only from Anamalais, Nilgiris and Palnis (Fig. 3). It grows on mossy tree-trunks and on shaded, wet rocks at higher altitudes (above 1000 m).

Notes : Kunze (1851) based Nilgiri specimen for his *G. attenuata*. It is characterised by uniformly dark-brown palea, elongate receptacle and sporangia intermixed with long bristles. Sledge (1960) described the sporangia of *G. attenuata* as non-setose, however, in our collections from the Nilgiris, setae are very rarely seen on the sporangia.

Specimens examined : TAMIL NADU : Coimbatore Dt., Anamalais, *Madhusoodanan* 43872 (CALI). Nilgiri Dt., Naduvattam, *Santhosh Nampy* 45162; way to Pykara, *Santhosh Nampy* 51993 (CALI).

Grammitis austroindica Parris inedit.

Type : South India, Nilgiris, *Beddome s. n.* (holotype, BM).

Rhizome long creeping, 0.8 mm in diameter, densely paleate. Palea ovate-lanceolate, *ca.* 2×0.8 mm, entire, uniformly pale-brown. Fronds simple, 2-4 mm apart, lanceolate, 3.5-8 × 0.4-0.9 cm, entire, acute at apex; stipe 2-4 mm long. Hairs on stipe 1.3-2 mm long, red-brown, hairs scattered to frequent on stipe, occasionally on the costa adaxially and lamina; 1 or 2-celled glandular hairs occasionally on the costa abaxially, occasional to sparse on either side of the lamina; short red-brown hairs 0.5-0.7 mm very sparsely on margin. Sori one row on either side of the midrib, round, nearer to the margin than costa, without acicular bristles. Sporangia non-setose.

Distribution and ecology : *G. austroindica* is endemic to South India and known only from Nilgiris. (Fig. 3).

Notes : Sledge (1960) was the first to consider this as a new species, but has not published it. While treating the Polypodiaceae and Grammitidaceae of Ceylon, Sledge (1960 : 149) commented "a Nilgiri gathering of his in Herb. Brit. Mus. labelled *P. parasiticum* with markedly creeping rhizome, long reddish hairs on the stipes and broad glabrous fronds with submarginal sori appears to represent an undescribed species" and noted on the herbarium specimen as" *Grammitis nov. spec.*, not *G. attenuata* Kze., no *mediale* Bak." It is easily recognisable from other South Indian species of *Grammitis* in its long creeping rhizome, uniformly pale-brown palea and glabrous fronds with long reddish hairs on the stipe. Parris (Pers. Comm.) is describing this as a new species.

Specimen examined : TAMIL NADU : Nilgiri Dt., *s. loc.*, *Beddome s. n.* (BM).

Grammitis pilifera Ravi & Joseph, Jour. Bombay Nat. Hist. Soc. 76 : 348. 1979; Nampy & Madhusoodanan, Indian Fern Jour. 9 : 206. 1992.

Type : South India, Ponmudi, *Ravi* 5711A (holotype, CAL; isotype, MH).

Polypodium parasiticum sensu Beddome, Handb. Ferns Brit. India 302. 1883, in part, non Mettenius, 1857.

Grammitis medialis sensu Nayar & Kaur, Comp. Bedd. Handb. Ferns Brit. India 75. 1974; Manickam & Irudayaraj, Pterid. Fl. W. Ghats 337. pl. 255. 1992, in part, non *Polypodium mediale* Bak., 1874.

Grammitis attenuata sensu Nayar & Geevarghese, Fern Fl. Malabar 348. 1993, in part, non Kunze, 1851.

(Fig. 5)

Grammitis pilifera var. *munnarensis* Raju, Shanavas, Santhosh & Thomas, Jour. Econ. Tax. Bot. 20 : 697-698. 1996.

Rhizome erect, short, 1-2 mm in diameter, densely paleaceous all over; solenostelic, stele sheathed by broad band of thick walled cells. Palea peltate with sub-basal attachment, 2 × 0.7 mm, lanceolate or broadly lanceolate, uniformly pale-brown except a few central dark-brown cells, apex blunt, glandular, margins entire. Fronds simple, 50 × 4 mm including the stipe; stipe terete, 5 × 0.5 mm, non-articulated, velutinous, short (0.4 mm long) hairy; lamina linear-oblong or oblong-lanceolate, margins entire, apex blunt, base gradually decurrent, herbaceous to sub-coriaceous, revolute on drying, adaxial surface frequently short (0.6 mm long) hairy, abaxial surface with a few hairs; hairs simple, acicular, reddish-brown. Venation distinct, veins simple or forked towards the soral region, thick and dark coloured upto midway. Sori in one row on either side of the midrib, round, 1.5 mm in diameter, median or supra-median, distinct, receptacle ovate, without bristles; paraphyses 1-4-celled, uniseriate. Sporangia setose or occasionally non-setose; setae 1-3 per sporangium; stalk 1-rowed, one time longer than the capsule. Capsule round; annulus 10-12-celled. Spores 43 × 43 µm, trilete-globose, verrucate.

Distribution and ecology : *G. pilifera* is endemic to South India (Fig. 3). However, this species has a more wider distribution in South India when compared to *G. attenuata*. It grows on mossy tree trunks or on wet rocks in and along the mountain streams usually above 800 m altitude.

Notes : *G. pilifera* was described by Ravi and Joseph (1979) based on a specimen from Ponmudi, Kerala. This species is closely related to *G. medialis* (Bak.) Ching. Manickam and Irudayaraj (1992) treated the South Indian specimens belonging to this species as *G. medialis*. However, the present study have shown that, the South Indian specimens that have been discussed under *G. medialis* actually pertains to *G. pilifera*. It differs from *G. medialis* in having fronds hirsute abaxially and having peltate palea with sub basal attachment. Recently, we have studied all the specimens from South India usually referred to *G. medialis* available at BM & K and found that they belong to *G. pilifera*. Dr. Parris (Pers. Comm. dated 27. Nov. 1994) also states "I have seen no material matching the type of *Grammitis medialis* from South India...".

Specimens examined : KARNATAKA : Chikmagalur Dt., Bababudan hills, *Santhosh Nampy* 48840 (CALI). KERALA : Kozhikode Dt., way to Vellarimala, *Santhosh Nampy* 48861 (CALI). Thiruvananthapuram Dt., way to Agasthiamala, *Santhosh Nampy* 48627; way to Ponmudi, *Santhosh Nampy* 48881; Ponmudi, *Pradeep* 43702 (CALI). Wynad Dt., way to Chembra peak, *Santhosh Nampy* 45912; Chembra, *Geevarghese & Madhusoodanan* 29286; *Nayar* 21314; Periya, *Santhosh Nampy* 45968 (CALI). TAMIL NADU : Kattabomman Dt., Kakachi river bank, *Santhosh Nampy* 45123; way to

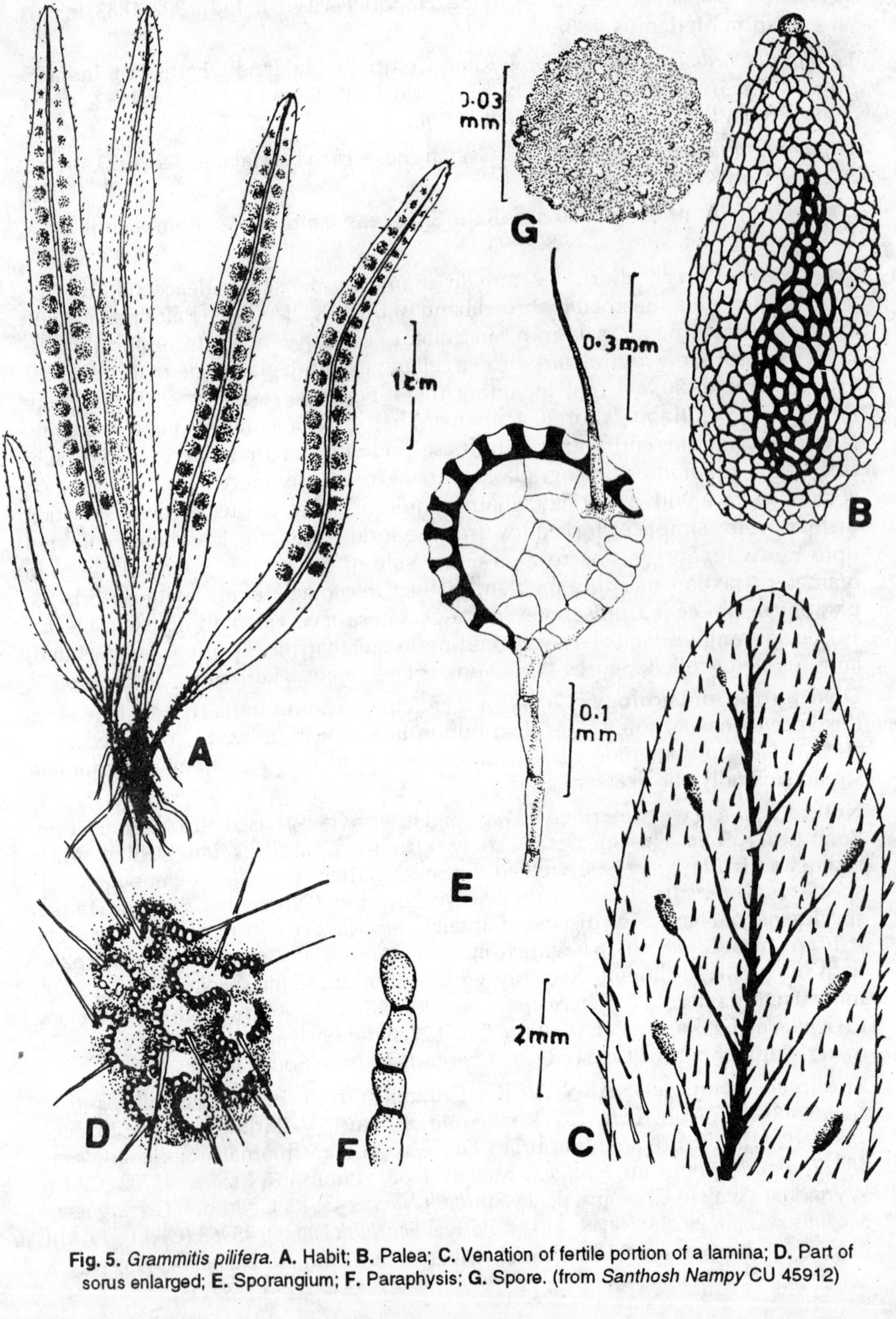

Fig. 5. *Grammitis pilifera*. A. Habit; B. Palea; C. Venation of fertile portion of a lamina; D. Part of sorus enlarged; E. Sporangium; F. Paraphysis; G. Spore. (from *Santhosh Nampy* CU 45912)

Kothayar, *Santhosh Nampy* 48662; Nalumukku, *Santhosh Nampy* 51994; Upper Kothayar, *Madhusoodanan* 44857, *Santhosh Nampy* 45193 (CALI).

PROSAPTIA Presl

The genus *Prosaptia* is closely related to *Ctenopteris* Blume and *Grammitis* Swartz. Copeland (1947) distinguished *Prosaptia* from the former by its deeply sunken sori. However, Holttum (1954) merged *Prosaptia* under *Ctenopteris* because "there is no really sharp division between these and typical species of *Ctenopteris*". Recently, Parris (1991) reduced both these genera under *Grammitis* because of their gradual and continuous intergradation in many characters, but she admittedly stated that "the type species of *Prosaptia* and its relatives may appear distinct enough to warrant generic separation...". In the light of the present investigation, we prefer to keep *Prosaptia* as separate from *Grammitis* and *Ctenopteris* , though they resemble each other in many respects. The ciliate rhizome scales and deeply immersed sori in bilipped cavities not known anywhere in this group is sufficient for its generic separation. The name 'Prosaptia' possibly refers to its immersed sori in the margins of the frond.

Prosaptia Presl, Tent. Pterid. 165. 1836; Beddome, Ferns South India 16. 1864 & Handb. Ferns Brit. India 56. 1883; Copeland, Gen. Fil. 220. 1947; Devol in Li *et al.* (eds.), Fl. Taiwan 1 : 225. 1975; Manickam & Irudayaraj, Pterid. Fl. W. Ghats 339. 1992.

Type species : *Prosaptia pinnatifida* Presl

Rhizome short creeping, densely paleaceous. Palea lanceolate, dark-brown, margins ciliate. Fronds tufted, pinnatifid with contiguous bases; stipe short, non-articulated, hairy; pinnules elliptic or lanceolate, basal pinnules triangular and decurrent nearly to the base, coriaceous or sub-coriaceous. Venation pinnate-free. Sorus immersed in marginal or dorsal cavities; paraphysis absent. Annulus 12-14-celled. Spores trilete, spherical, chlorophyllous.

Gametophyte : Not reported

Chromosome numbers : x = 36 or 37 (Manton & Sledge, 1954).

Distribution and ecology : *Prosaptia* is distributed from South India to Polynesia and New Guinea. Many of its species occur either as epiphytes or lithophytes in medium to higher elevations (500 m or above) in exposed or shaded situations.

Key to the species

1a. Lamina pinnately lobed to *ca.* 2-3 mm from the midvein..*P. alata*

1b. Lamina pinnate or pinnately lobed to less than I mm from the midvein.......................2

2a. Fronds coriaceous; sori immersed in marginal or apical bilipped cavities...*P. contigua*

2b. Fronds sub-coriaceous; sori immersed abaxially in elliptical cavities.....................*P. obliquata*

Prosaptia alata (Bl.) C. Chr., Ann. Jard. Bot. Buit. 2, 5 : 127. 1905; Ching, Bull. Fan Mem. Inst. Biol. 10 : 241. 1941; Sledge, Bull. Brit. Mus. (Nat. Hist.) Bot. 2(5) : 157. 1960; Satija & Bir, Polyp. Ferns India 4. 1985.

Davallia alata Blume, Enum. P1. Jav. 230. 1828.

Type : Java, *Zippel s. n.* (lectotype, L).

Davallia emersonii Hook. & Grev., Ic. Fil. 1 : t. 105. 1829.

Prosaptia emersonii Presl, Tent. Pterid. 166. 1836; Beddome, Ferns South India 6. t. 20. 1864 & Handb. Ferns Brit. India 56. t. 27. 1883; Nayar & Kaur, Comp. Bedd. Handb. Ferns Brit. India 17. 1974.

Ctenopteris alata (Bl.) Holttum, Fern F1. Malaya 2 : 232. 1954.

Rhizome short creeping, 2 mm wide without scales, densely paleaceous. Palea basifixed, ovate, narrowly-ovate or linear-lanceolate, 4 × 1 mm, clathrate, glandular or ciliate at apex, hairy abaxially and on margins. Fronds articulated, 1 mm apart, 21.5-30 × 1.9-2.4 cm, narrowed gradually to the base and apex; stipe 0.4-1.2 cm; lamina pinnately lobed to *ca.* 2-3 mm from the midvein; pinnules linear-oblong, triangular at the base, coriaceous. Simple, 0.2-0.5 mm long, red-brown hairs sparse to frequent on the midrib, stipe, margin and lower surface of the lamina. Venation pinnate-free. Sori 2-9 on each pinnules, pustular abaxially; soral cavities oblique to the edge and distinctly separated from it. Annulus 11 or 12-celled. Spores trilete, tetrahedral, chlorophyllous, verrucate.

Distribution and ecology : *P. alata* is very rare in South India (Fig. 3), being only once collected by Beddome from Anamalais and Palghat hills. It is also distributed in North India, Sri Lanka, Thailand, Peninsular Malaysia, Sumatra, Java, Borneo, Sulawesi, Philippines and Polynesia.

Notes : *P. alata* is easily recognisable from other species of this genus in South India by its less deeply lobed, coriaceous fronds and the soral cavities oblique to the edge of the pinnules.

Specimens examined : KERALA : Palakkad Dt., *s. loc.*, *Beddome s. n.* (BM). TAMIL NADU : Coimbatore Dt., Anamalais, *Beddome s. n.* (BM, K).

Prosaptia contigua (Forst. f.) Presl, Tent. Pterid. 166. 1836; Beddome, Ferns South India 6. t. 19. 1864 & Handb. Ferns Brit. India 56. 1883; Sledge, Bull. Brit. Mus. (Nat. Hist.) Bot. 2(5) : 157. 1960; Nayar & Kaur, Comp. Bedd. Handb. Ferns Brit. India 17. 1974; Satija & Bir, Polyp. Ferns India 4. 1985; Manickam & Irudayaraj, Pterid. Fl. W. Ghats 341. pl. 258. 1992.

Trichomanes contiguum Forst. f., Flourl. Ins. Austr. Prodr. 84. 1786.

Davallia contigua (Forst. f.) Swartz, Synops. Fil. 130. 1806.

Polypodium contiguum (Forst. f.) J. Sm., Jour. Bot. 3 : 394. 1841.

Ctenopteris contigua (Forst. f.) Holttum, Fern Fl. Malaya 2 : 230. 1954.

(Figs. 6, 8a, 14 a)

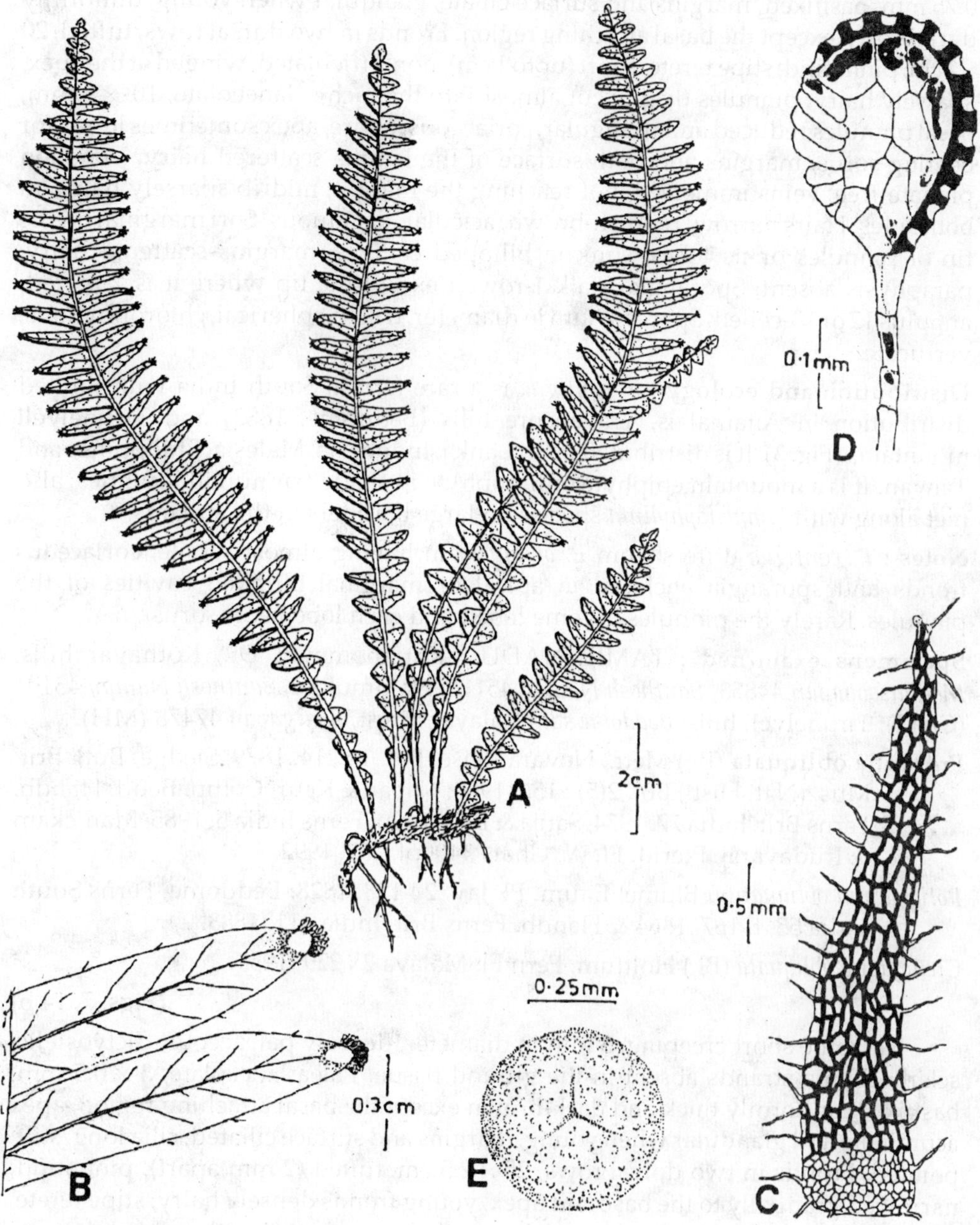

Fig. 6. *Prosaptia contigua.* **A.** Habit; **B.** Venation of fertile portion of a lamina; **C.** Palea; **D.** Sporangium; **E.** Spore. (from *Santhosh Nampy* CU 45116)

Rhizome short creeping, 3 mm in diameter, densely paleaceous all over; dictyostelic, sclerenchyma strands absent in the ground tissue. Palea lanceolate, 3 × 0.75 mm, basifixed, margins and surface ciliate, glandular when young, uniformly dark-brown except the basal attaching region. Fronds in two dorsal rows, tufted, 20 × 2 cm, pinnatifid; stipe terete, short (upto 1 cm), non-articulated, winged at the apex, sparsely hairy; pinnules deeply cut almost into the rachis, lanceolate, 10 × 3 mm, basal pinnules reduced and triangular, coriaceous, entire, apex sometimes lobed for bearing sorus, margins and both surface of the lamina scattered hairy. Venation pinnate-free, veins immersed, not reaching the margin, midrib sparsely hairy on both sides. Hairs narrow, reddish-brown, acicular, deciduous. Sori marginal on the tip of pinnules or its lobes, sunk in bilipped cavities, margins scattered hairy; paraphysis absent. Sporangial stalk 1-rowed except the tip where it is 2-rowed, annulus 12 or 13-celled. Spores 49 µm in diameter, trilete, spherical, chlorophyllous, verrucate.

Distribution and ecology : *P. contigua* is a rare fern in South India with limited distribution in Anamalais, Travancore hills (Beddome, 1883) and Tirunelveli mountains (Fig. 3). It is distributed in Sri Lanka, Indonesia, Malesia, Philippines and Taiwan. It is a mountain epiphyte or lithophyte in shaded or humid localities; also met along with *Hymenophyllum* species on large boulders in the streams.

Notes : *P. contigua* differs from *P. obliquata* in having almost sessile, coriaceous fronds and sporangia enclosed in apical or marginal bilipped cavities of the pinnules. Rarely the pinnules become lobed and each lobe bears sorus.

Specimens examined : TAMIL NADU : Kattabomman Dt., Kothayar hills, *Madhusoodanan* 44853, *Santhosh Nampy* 45116; Nalumukku, *Santhosh Nampy* 45191 (CALI); Tirunelveli hills, *Beddome s.n.*; Valayar forest, *Bhargavan* 47478 (MH).

Prosaptia obliquata (Bl.) Mett., Novara Reise. Bot. 1 : 214. 1879; Sledge, Bull. Brit. Mus. (Nat. Hist.) Bot. 2(5) : 156. 1960; Nayar & Kaur, Comp. Bedd. Handb. Ferns Brit. India 77. 1974; Satija & Bir, Polyp. Ferns India 5. 1985; Manickam & Irudayaraj, Pterid. Fl. W. Ghats 340. pl. 257. 1992.

Polypodium obliquatum Blume, Enum. Pl. Jav. 2 : 128. 1828; Beddome, Ferns South India 55. t. 167. 1864 & Handb. Ferns Brit. India 311. 1883.

Ctenopteris obliquata (Bl.) Holttum, Fern Fl. Malaya 2 : 229. 1954.

(Figs. 7, 14 b)

Rhizome short creeping, 3 mm in diameter, densely paleaceous; dictyostelic, sclerenchyma strands absent in the ground tissue. Palea lanceolate, 3 × 0.5 mm, basifixed, uniformly thick and dark-brown except the basal attaching region, apex acuminate and glandular when young, margins and surface ciliated; cilia long, stiff, pointed. Fronds in two dorsal rows, 25 × 2.5 cm, tufted (2 mm apart), pinnatifid; narrowed gradually to the base and apex, young fronds densely hairy; stipe terete, 0.6-3 cm × 1 mm, gradually winged above, densely covered with reddish-brown, acicular, deciduous hairs, non-articulated; pinnules deeply cut almost to the rachis, elliptic or lanceolate, basal ones triangular and reduced, 12-16 × 2-3 mm, broadly adnate to the costa, slightly ascending, entire, recurved on drying, sub-coriaceous, a very few hairs scattered on the margins and both surfaces. Venation pinnate-free,

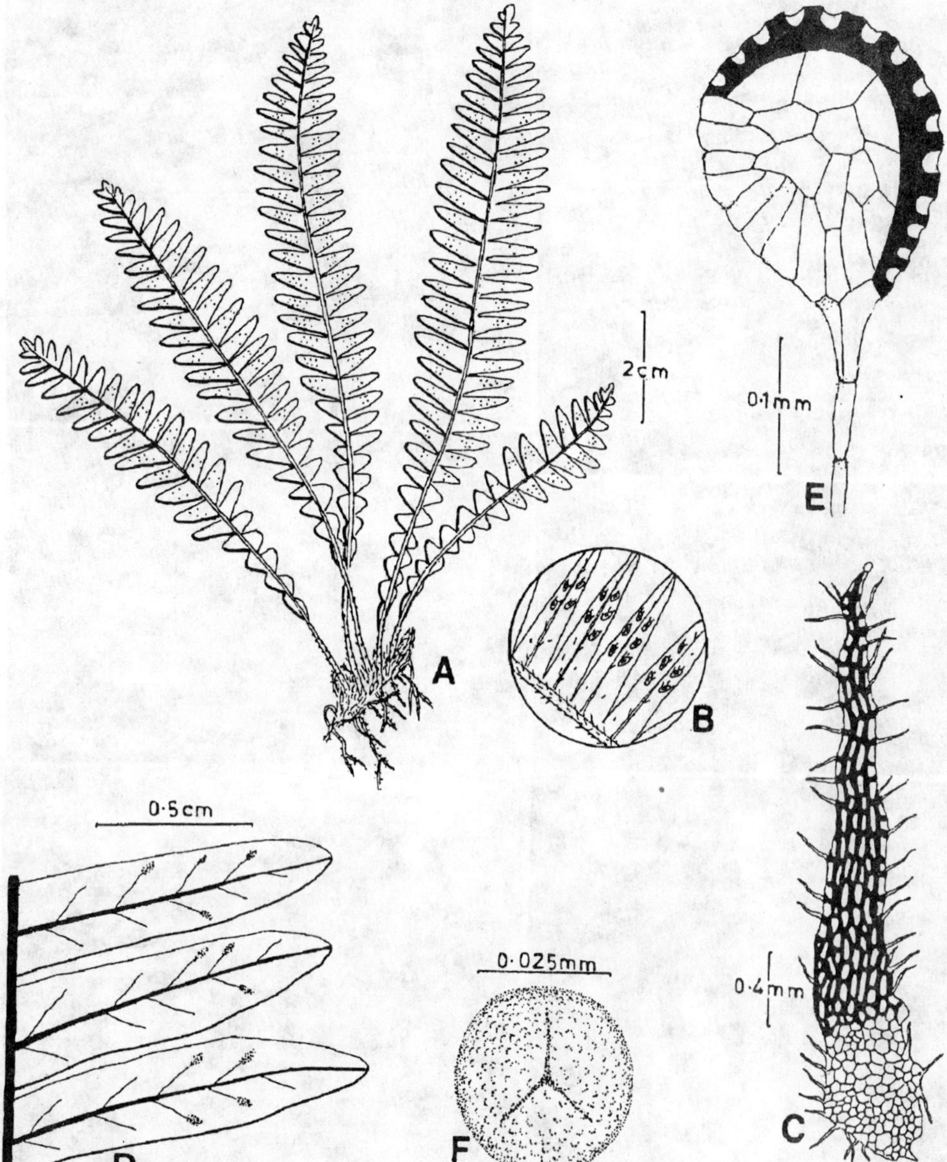

Fig. 7. *Prosaptia obliquata.* **A.** Habit; **B.** Part of fertile frond enlarged; **C.** Palea; **D.** Venation of fertile portion of a lamina ; **E.** Sporangium; **F.** Spore. (from *Santhosh Nampy* CU 45120)

Fig. 8. Habit of South Indian polypodioid ferns. **a.** *Prosaptia contigua;* **b.** *Loxogramme cuspidata;* **c.** *L. chinensis;* **d.** *Lepisorus nudus;* **e.** *L. amaurolepidus;* **f.** *Drynaria quercifolia*

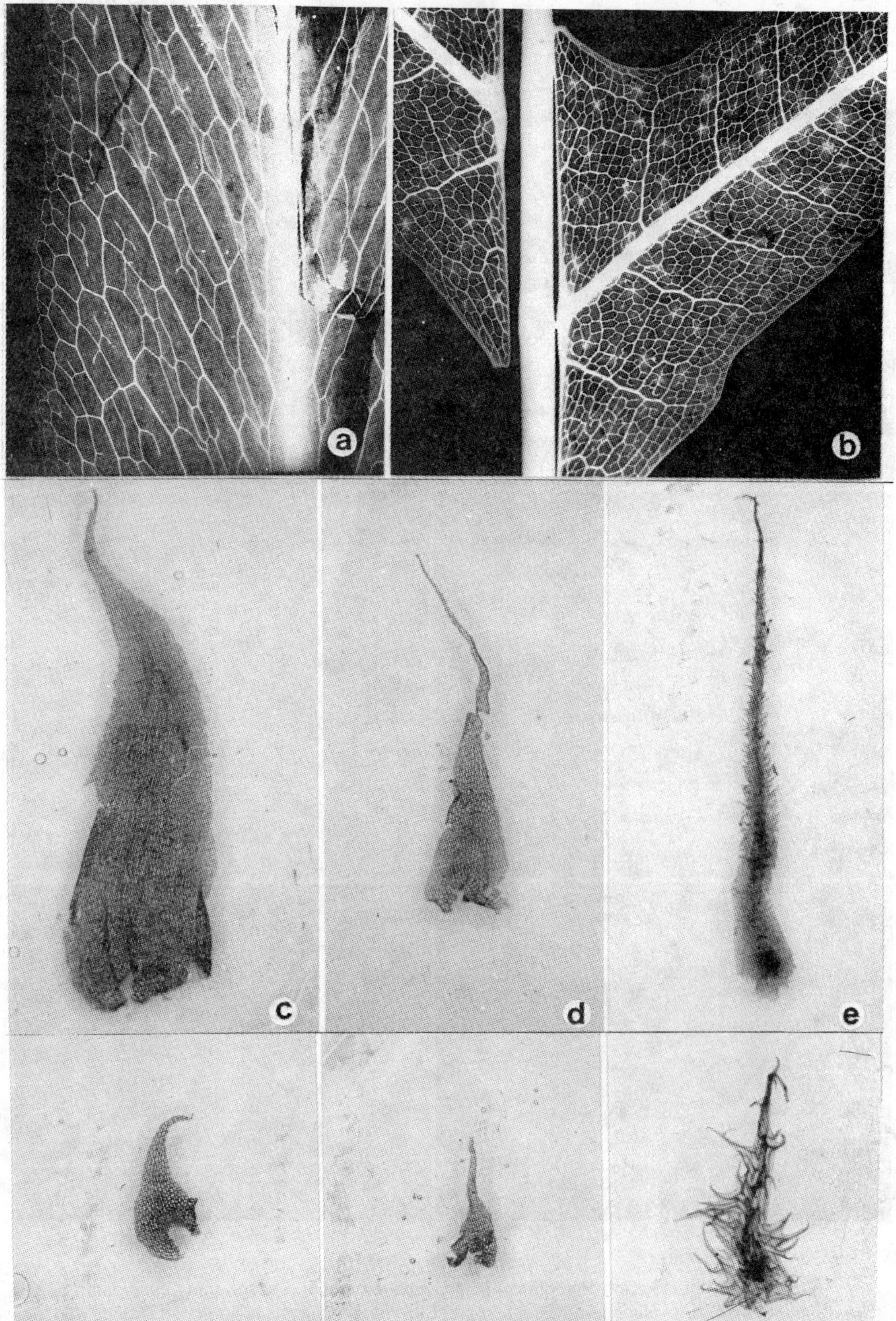

Fig. 9. Venation and palea of polypodioid ferns. **a.** *Loxogramme cuspidata* × 1.5; **b.** *Drynaria quercifolia* × 1.5; **c-h.** Palea with LM; **c, f.** *Loxogramme cuspidata* × 4.5; **d, g.** *L. chinensis* × 4.5; **e.** *Drynaria quercifolia* × 4.5; **h.** *D. quercifolia*, palea from lamina × 18.75.

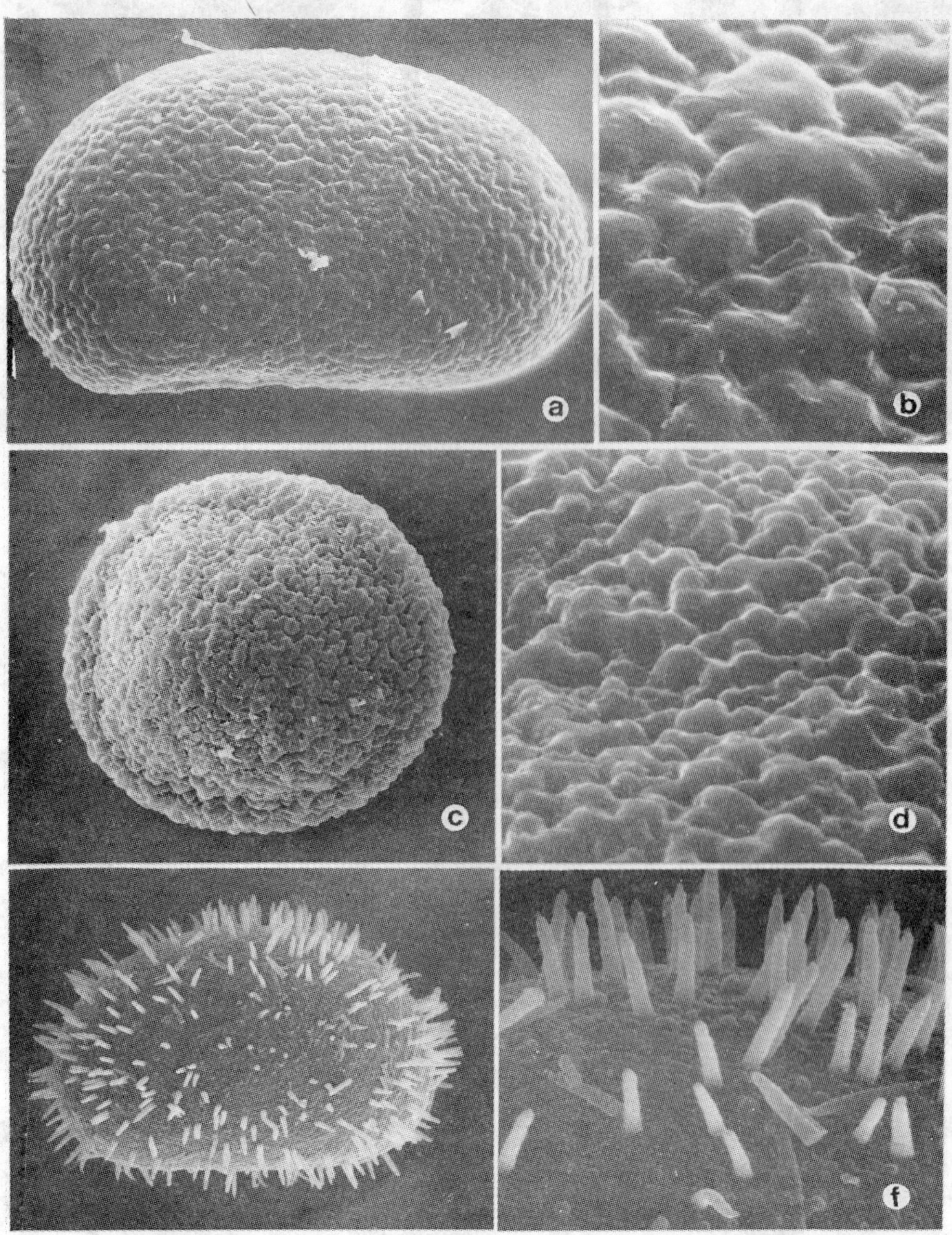

Fig. 10. SEM photograph of spores of polypodioid ferns. **a, b.** *Loxogramme cuspidata* (Kew Distr. 1079); **c, d.** *L. chinensis (Mann s. n.)*; **e. f.** *Drynaria quercifolia (Santhosh Nampy 48615)*. **a, c, e.** Lateral view; **b, d, f.** A part enlarged. **a, e.** × 1125; **c.** × 1500; **b, d.** × 5625; **f.** × 3750

veins immersed, not reaching the margin. Sori elliptic, 1 × 0.5 mm, supra-median, produced at the vein endings, sunken in oblique-elliptic cavities, pustular adaxially, regularly arranged on either side of the main vein, usually 2-8 per pinnule; non-paraphysate. Annulus 13 or 14-celled, sporangial stalk 1-rowed except at the tip where it is 2-rowed. Spores 19 μm in diameter, trilete, spherical, chlorophyllous, verrucate.

Distribution and ecology : *P. obliquata* is rare in South India (Fig. 3). It is also distributed in Sri Lanka, Indo-China, Sumatra to New Guinea. It is a mountain epiphyte or lithophyte in dense, humid forests of higher altitudes (above 1100 m).

Notes : *P. obliquata* is characterised by its sub-coriaceous fronds, hairy stipe and rachis, elliptic, deeply sunken sori dorsally on the pinnules.

Specimens examined: KERALA: Idukki Dt., Maddalamkotti near Elatheri, *Mohanan* 72804 (MH). TAMIL NADU : Kattabomman Dt., Kothayar hills, *Madhusoodanan* 44854, *Santhosh Nampy* 45120; Nalumukku, *Santhosh Nampy* 45192 (CALI); Tirunelveli hills, *Beddome s.n.* (MH). Ramanathapuram Dt., Sethur hills, *Srinivasan* 65988 (MH).

LOXOGRAMMACEAE Ching ex Pic. Serm.

The family Loxogrammaceae was established by Ching (1940) as monotypic. He defined the family as "...leaves,...not articulate to rhizome,...venation regularly reticulate with oblique elongate hexagonal areoles; as a rule without free included veinlets...Spores bilateral,..." and considered the family as allied to Antrophyaceae Ching and Vittariaceae Presl. Loxogrammaceae as established by Ching (1940) was invalid, since he failed to provide a latin diagnosis for the family. Subsequently, Pichi Sermolli (1974) validated the name. Many authors of major classification (Wilson, 1959; Nayar, 1970; Pichi Sermolli, 1977) have followed Ching (*l.c.*) in maintaining Loxogrammaceae as a distinct family while Wagner (1961) kept it as a subfamily of Polypodiaceae. However, several others (Copeland, 1947; Holttum, 1949; Pichi Sermolli, 1958; Hennipman *et al.*, 1991) placed it under Polypodiaceae.

Pichi Sermolli (1974), however, has questioned Ching's (1940) view on the affinity of Loxogrammaceae to Antrophyaceae and Vittariaceae. According to him, Vittariaceae (including *Antrophyum* and related genera) have a chromosome number x = 15 while Loxogrammaceae have x = 35, 36 or 37 and considered the latter as more close to Grammitidaceae and Polypodiaceae. As currently understood, the family includes two genera, namely *Anarthropteris* Copeland and *Loxogramme* (Bl.) Presl, the former is monotypic and is rather confined to New Zealand while the latter includes about 33 species distributed in Asia, Africa and Central America.

Loxogrammaceae Ching ex Pic. Serm., Webbia 29 : 11. 1974.

Type genus : *Loxogramme* (Bl.) Presl

Rhizome short or long creeping, paleaceous; dictyostelic, having sclerenchyma strands in the ground tissue. Palea entire, concolorous. Fronds in two dorsal rows on the rhizome, simple, monomorphic to dimorphic; stipe very short, non-articulate; lamina linear-lanceolate, entire, herbaceous. Venation reticulate with oblique elongate hexagonal areoles, free included veinlets none. Sori oblique or sub-parallel to the costa, exindusiate, elongate. Sporangial stalk 3-rowed at the capsular base; annulus *ca.* 12-celled. Spores monolete or trilete (in some cases trilete spores are found intermixed with monolete as well as intermediate forms within the same sporangium), exine granulose.

Notes : In South India, the family is represented by the genus *Loxogramme* with two species.

LOXOGRAMME (Bl.) Presl

The name 'Loxogramme' was derived from 'loxos' meaning oblique, and 'gramme' meaning line; referring to its oblique orientation of the sorus. Many species that are currently placed under the genus *Loxogramme* were earlier treated under *Antrophyum* sect. *Loxogramme* Blume. It was Presl (1836) who elevated Blume's (1828) sect. *Loxogramme* to the rank of a genus. The genus is characterised by anastomosing lateral veins, clathrate palea, stipes not functionally jointed to the rhizome, flaccid-leathery, glabrate lamina which roll revolutely or involutely on drying, elongate sori and chlorophyllous spores. The spores of this genus are unique in their diverse forms ranging from trilete-globoidal to monolete-ellipsoidal, and intermediate types are occasionally seen. There have been essentially no disputes concerning its generic boundary or rank eversince its establishment.

Loxogramme (Bl.) Presl, Tent. Pterid. 214. 1836; Beddome, Ferns South India 17. 1864 & Handb. Ferns Brit. India 392. 1883; C. Chr., Ind. Fil. 410. 1906; Ching, Sunyatsenia 5 : 233. 1940; Copeland, Gen. Fil. 217. 1947; Holttum, Fern Fl. Malaya 2 : 166. 1954; Nayar, Jour. Ind. Bot. Soc. 34 : 395-407. 1955; Tryon & Tryon, Ferns Allied Plants 743 . 1982; Satija & Bir, Polyp. Ferns India 6.1985; Price in Kubitzki (ed.), Fam. Gen. Vas. Plants 1 : 229. 1991; Manickam & Irudayaraj, Pterid. Fl. W. Ghats 313. 1992; Nampy & Madhusoodanan. Indian Fern Jour. 9 : 208. 1992.

Antrophyum sect. *Loxogramme* Blume, Fl. Jav. Filices 73. 1829.

Type species : *Loxogramme lanceolata* (Sw.) Presl

Rhizome short or long creeping, slender, often with masses of hairy roots, paleaceous; dictyostelic. Palea lanceolate, thin, clathrate, pale brown to blackish, concolorous, cells elongate. Fronds simple, shortly stipitate, not functionally articulated; lamina lanceolate, oblanceolate or linear-lanceolate, fleshy or flaccid, margins rolled involutely or revolutely on drying. Veins hidden, anastomosing, areoles oblique, elongated areoles with or occasionally without free included veinlets. Sori elongate, exindusiate, oblique to the costa, ends of sori often overlapping, with or without paraphysis. Sporangia with 13 or 14-celled annulus. Spores globose-tetrahedral or monolete-bilateral or both, chlorophyllous.

Gametophyte : The mature prothallus of *Loxogramme* is of the ribbon-like type, and often profusely branched so that the thallus appears as an irregular plate-like mass (Nayar, 1968).

Chromosome numbers : n = 35, 36, 70 ; 2n = 72 (cf. Tryon & Tryon, 1982).

Distribution and ecology : *Loxogramme* has about 33 species mainly confined to Malesia and tropical and warm temperate continental Asia (Price, 1991). Eight species are recorded for India of which two, namely, *L. chinensis* Ching and

L. cuspidata (Zenker) Price are distributed in South India. *Loxogramme* prefers shaded, humid conditions in the forests and grows as tree trunk epiphyte or petrophyte.

Key to the species

1a. Margins of lamina curl involutely on drying ; spores always monolete............*L. cuspidata*

1b. Margins of lamina curl revolutely on drying ; spores mostly trilete...............*L. chinensis*

Loxogramme chinensis Ching, Sinensia 1 : 13. 1929 & Ic. Fil. Sin. 1 : 46. 1930; Satija & Bir, Polyp. Ferns India 6. 1985; Madhusoodanan & Nampy, Indian Fern Jour. 11 : 56-59. 1994.

Loxogramme lanceolata sensu Beddome Ferns South India 17. t. 51. 1864 & Handb. Ferns Brit. India 392. 1883; Nampy & Madhusoodanan, Indian Fern Jour. 9 : 209. 1992, non Presl, 1836.

Loxogramme parallela sensu Nayar & Kaur, Comp. Bedd. Handb. Ferns Brit. India 94. 1974; Chandra & Kaur, Nomen. Guide Bedd. Ferns South India & Ferns Brit. India 6. 1987; Manickam & Irudayaraj, Pterid. Fl. W. Ghats 315. pl. 240. 1992, non Copeland, 1905.

(Figs. 8 c, 11, 14 c)

Rhizome short creeping, *ca.* 2 mm in diameter, densely paleaceous. Palea lanceolate to ovate-lanceolate, *ca.* 5 × 1.5 mm, basifixed, clathrate, margins entire, gland-tipped. Fronds sessile or sub-sessile, 9-15 × 0.7-1.2 cm, 4-6 mm apart, the base clockwisely twisted a little, simple, not functionally articulated; lamina linear-lanceolate to oblanceolate, margins smooth or slightly wavy, base gradually decurrent into a narrow wing ending to the rhizome, apex acute or gradually tapering, glabrous, thick, herbaceous, curl revolutely (towards the soral side) on drying. Venation reticulate, midrib slightly raised on both sides, other veins not prominent externally, excluded and included veins are occasionally seen towards the apex of the lamina. Sori linear, oblique to sub-parallel, 6 × 2 mm, never overlapping with the ends of sori below and above; paraphyses long, filamentous, rarely branched. Sporangium long-stalked, stalk 2-rowed except at the apex where it is 3-rowed, annulus 13-celled. Spores 64 × 54 µm, trilete, tetrahedral, verrucate, greenish-yellow; monolete, bilateral forms also present.

Distribution and ecology : *Loxogramme chinensis* has a limited distribution in southern (Fig. 3) and northern India. It also occurs in South West China, Upper Burma, northern Thailand and Tonkin. The plant prefers humid situations and grows on moist, mossy tree trunks in forests. During dry months, leaf margins roll revolutely to minimise water loss.

Notes : *Loxogramme chinensis* Ching is a widely distributed and frequently misunderstood taxon (Madhusoodanan & Nampy, 1994). This species is closely allied to *L. lanceolata* Presl and *L. parallela* Copeland. Many authors (Beddome, 1864, 1883; Nayar & Kaur, 1974; Satija & Bir, 1985) treated the South Indian specimens belonging to this species as *L. lanceolata,* while others (Sledge, 1960 ; Chandra & Kaur,

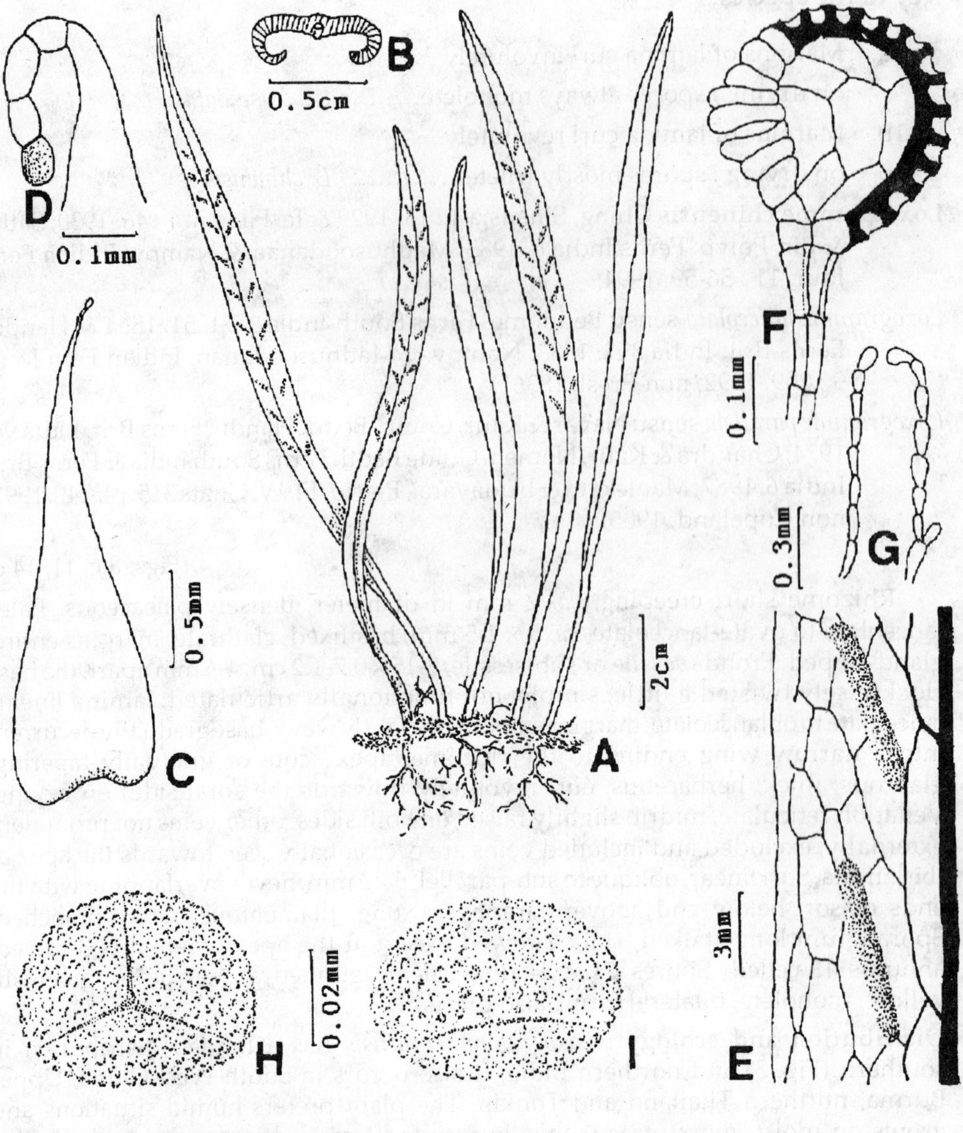

Fig. 11. *Loxogramme chinensis.* **A.** Habit; **B.** Frond c. s. showing volution; **C.** Palea; **D.** Apex of palea enlarged; **E.** Venation of fertile portion of a lamina; **F.** Sporanguim; **G.** Paraphyses; **H, I.** Spores. (from *Santhosh Nampy* CU 45159)

1987; Manickam & Irudayaraj, 1992) considered it as *L. parallela.* However, our studies on this genus in Peninsular India has shown that, the South Indian specimens that has been discussed either under the name *L. lanceolata* or *L. parallela* by all the earlier authors, actually pertains to *L. chinensis* Ching. Some of our collections of this species from the Nilgiris were referred to Dr. Price, University of Michigan, U.S.A. After examining our specimens, he commented (Pers. Comm. dated 16 Sept. 1992) that "this species has been collected previously in the Nilgiris, including by Beddome, who illustrated it in Ferns of Southern India plate 51, but who wrongly named it *Loxogramme lanceolata.* I have examined the actual specimen used for the plate by Beddome, at the British Museum, and it is positively the same species as your specimen. Its correct name is *Loxogramme chinensis* Ching". It differs from the other two species in having clathrate rhizome scales, linear-lanceolate to oblanceolate fronds, margins of the fronds curling revolutely on drying and having mostly trilete spores.

Specimens examined : KERALA : Kozhikode Dt., Vellarimala, *Santhosh Nampy* 201 (Devagiri College). TAMIL NADU : Nilgiri Dt., Naduvattam, *Santhosh Nampy* 45159 (CALI), *Subramanyam* 10581, *Ellis* 34747, *Vajravelu* 46067 (MH); way to Ooty (near Naduvattam), *Santhosh Nampy* 45173; near Pykara, *Santhosh Nampy* 45184 (CALI).

Loxogramme cuspidata (Zenker) Price, Amer. Fern Jour. 74 : 61. 1984; Nampy & Madhusoodanan, Jour. Econ. Tax. Bot. 19(3) : 741-744. 1995.

Grammitis cuspidata Zenker, Pl. Ind. 1 : t. 2. 1835; C. Chr., Ind. Fil. 331. 1906.

Type : South India (near Ootacamund), Rev. *B. Schmid s.n.* (isotypes, BR, E, L, P).

Loxogramme involuta sensu Beddome, Ferns South India 17. t. 50. 1864 & Handb. Ferns Brit. India 393. 1883; Sledge, Bull. Brit. Mus. (Nat. Hist.) Bot. 2(5) : 142. 1960; Nayar & Kaur, Comp. Bedd. Handb. Ferns Brit. India 94. 1974; Satija & Bir, Polyp. Ferns India 7. 1985; Manickam, Fern Fl. Palni Hills 150. 1986; Chandra & Kaur, Nomen. Guide Bedd. Ferns South India & Ferns Brit. India 6. 1987; Manickam & Irudayaraj, Pterid. Fl. W. Ghats 314. pl. 239. 1992; Nampy & Madhusoodanan, Indian Fern Jour. 9 : 209. 1992, non Presl, 1836.

(Figs. 8 b, 12, 13, 14 d)

Rhizome long creeping, *ca.* 1 cm in diameter, densely paleaceous; dictyostelic, meristeles 6 or 7, ground tissue parenchymatous. Palea ovate-lanceolate, 8.5 × 2 mm, basifixed, clathrate, concolorous, with a long attenuate hair-like tip ending in a prominent gland cell, margins entire. Fronds 3-6 mm apart, oblanceolate, 32.5 × 2.8 cm, base gradually decurrent, apex acuminate; stipe short, non-articulate but a saucer-shaped abscission pad is produced by the fallen leaves, base darkened especially on the abaxial side; lamina simple, entire, thick, herbaceous, curl involutely on drying. Venation reticulate, midrib prominent on both surfaces, lateral veins closely placed and lose their identity towards the margins, areoles numerous, 1-2 rows of marginal areoles smaller, free veinlets occasionally present, vein endings not swollen. Sori restricted to the anterior half of the lamina, linear, distinct, 24 × 1.2 mm, parallel to the main lateral veins, the base and apex overlap with the ends of those below and above them; paraphysis uniseriate, ending in a swollen cell.

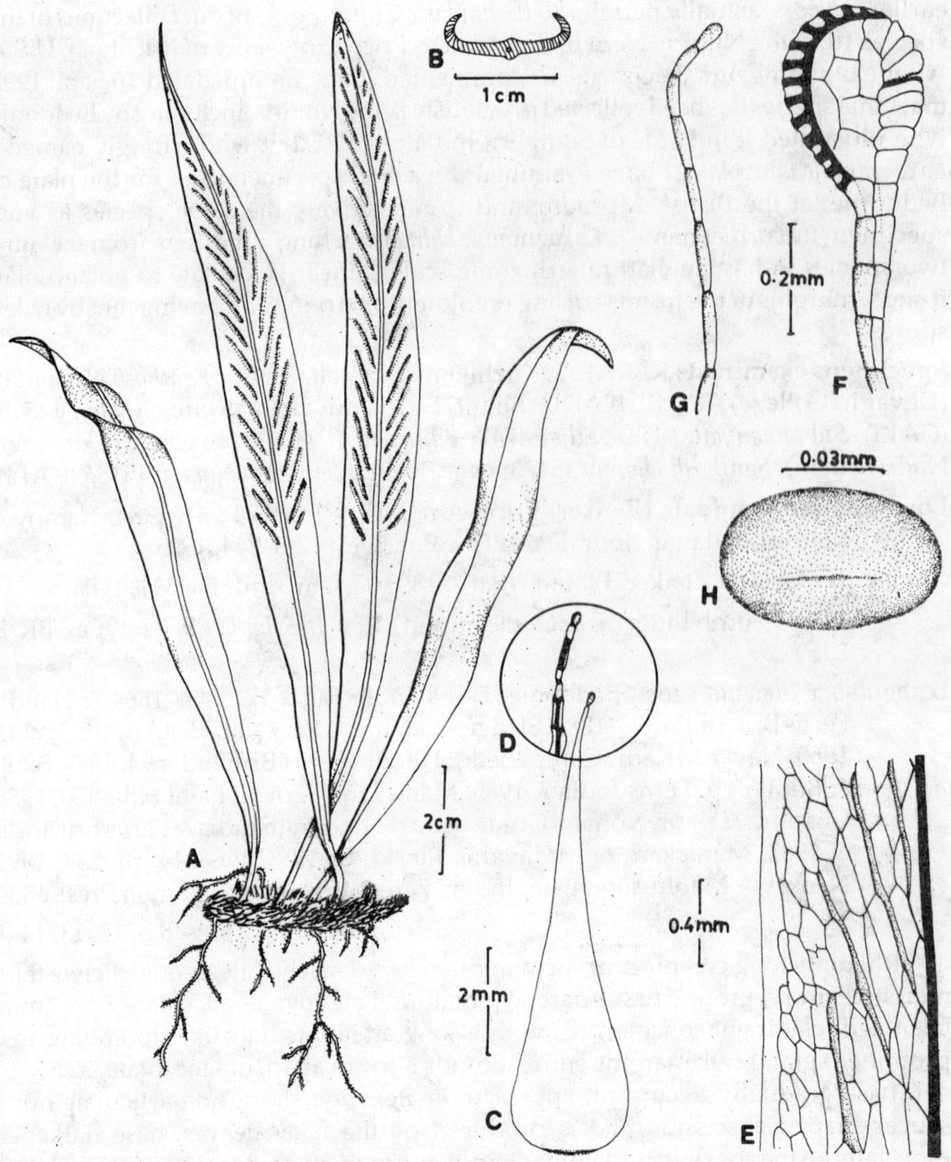

Fig. 12. *Loxogramme cuspidata.* **A.** Habit; **B.** Frond c.s. showing volution; **C.** Palea; **D.** Part of palea enlarged; **E.** Venation of fertile portion of a lamina; **F.** Sporangium; **G.** Paraphysis; **H.** Spore. (from *Santhosh Nampy* CU 45179)

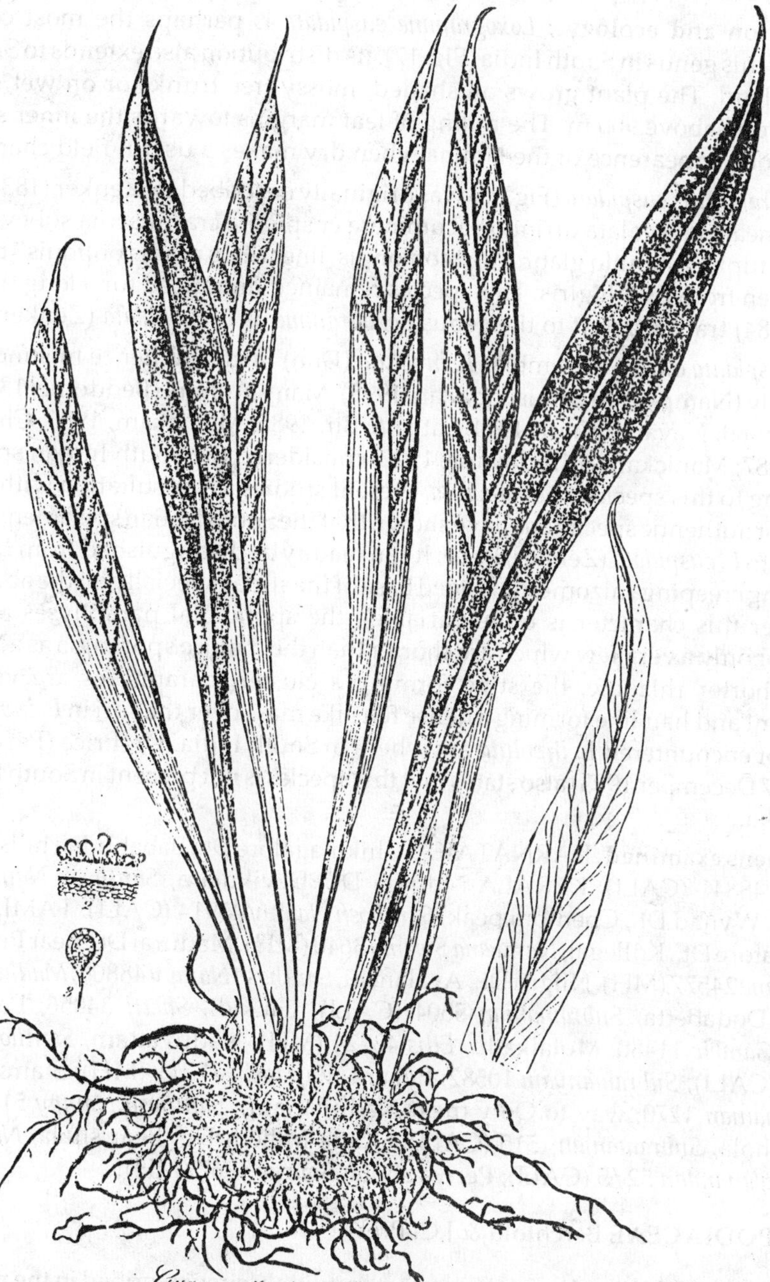

Fig. 13. *Loxogramme cuspidata* (Zenker) Price
(Type : B. *Schmid s.n.*, photocopy)

Sporangium golden-brown, annulus 13-celled, stalk long. Spores 66 × 44 μm, monolete, bilateral, golden-yellow, granulose.

Distribution and ecology : *Loxogramme cuspidata* is perhaps the most common species of this genus in South India (Fig. 17). Its distribution also extends to Sri Lanka and Thailand. The plant grows on shaded, mossy tree trunks or on wet rocks in dense forests above 900 m. The rolling of leaf margins towards the inner side and stramineous appearence of the lamina when dry makes a useful field character.

Notes : *Grammitis cuspidata* (Fig. 13) was originally described by Zenker (1835) as "G. fronde lineari-lanceolata utrinque acuminata cuspidata craffiuscula subevenia int-gerrima, ftipite fubnullo glabro, foris oblongis, linearibus planis obliquis" based on a specimen from the Nilgiris. This species remained unnoticed for a long time until Price (1984) transferred it to the genus *Loxogramme* as *L. cuspidata* (Zenker) Price.

L. cuspidata closely resembles *L. involuta* (Don) Presl and hence misunderstood frequently (Nampy & Madhusoodanan, 1995). Many authors (Beddome, 1864, 1883; Sledge, 1960; Nayar & Kaur, 1974; Satija & Bir, 1985; Manickam, 1986; Chandra & Kaur, 1987; Manickam & Irudayaraj, 1992) considered the South Indian specimens belonging to this species as *L. involuta*. Present studies in consultation with the type and other authentic specimens have shown that the South Indian specimens actually belongs to *L. cuspidata* (Zenker) Price. It can readily be distinguished from *L. involuta* by its long creeping rhizome, darkened base of the stipe especially on the abaxial side (however this character is quite variable), the sporangial paraphyses are either absent or only a very few which are shorter than the mature sporangia as against the much shorter rhizome, the stipe forming a cluster, paraphyses are very long, persistent and hair like forming a dense felt-like mass over the sori in *L. involuta*. We have not encountered *L. involuta* anywhere in South India. Dr. Price (Pers. Comm. dated 27 December 1993) also states that this species is not present in South India and Sri Lanka.

Specimens examined : KARNATAKA : Chikmagalore Dt., Bababudan hills, *Santhosh Nampy* 48841 (CALI). KERALA : Idukki Dt., Devikulam, *Santhosh Nampy* 48606 (CALI). Wynad Dt., Chembra peak, *Santhosh Nampy* 45914 (CALI). TAMIL NADU: Coimbatore Dt., Kollegal, *Narayana Swamy* 3640 (MH). Madurai Dt., near Poomparai, *Sebastiane* 24577 (MH). Nilgiri Dt., Avalanchi, *Santhosh Nampy* 48806, *Madhusoodanan* 51957; Dodabetta, *Subramanian* 5504 (CALI); Lakkidi, *Shetti* 34056; Laud's-vale shola, *Gamble* 11480; Mulaikadu, *Ellis* 43224 (MH); Naduvattam, *Santhosh Nampy* 51989 (CALI), *Subramanyam* 10582; Pykara, *Vajravelu* 77711 (MH); Staircase shola, *Subramanian* 1270; way to Ooty (near Naduvattam), *Santhosh Nampy* 51988; West Briar shola, *Subramanian* 515 (CALI). Quide Milleth Dt., Bear shola, *Nayar* 2722, *Sworrupanandan* 52/5 (CALI); Periya shola, *Bourne* 4967 (MH).

POLYPODIACEAE Berchtold & J.C. Presl

The family Polypodiaceae has been very widely circumscribed in the past. It was Ching (1940), who first used the family in a restricted sense, and this was followed by many later authors (for more details see page no. 11). Here the family is presented in a narrow sense excluding Cheiropleuriaceae, Dipteridaceae, Grammitidaceae and Loxogrammaceae but including Drynariaceae and Platyceriaceae. It is charac-

Fig. 14. Fronds of polypodioid ferns showing soral morphology and distribution. **a.** *Prosaptia contigua;* **b.** *P. obliquata;* **c.** *Loxogramme chinensis;* **d.** *L. cuspidata;* **e.** *Lepisorus nudus;* **f.** *Pleopeltis macrocarpa;* **g, h.** *Phymatosorus nigrescens;* **i.** *Pyrrosia lanceolata*

terised by usually epiphytic habit, having a perforated dictyostele, fronds articulated to phyllopodia, reticulate venation, sporangial stalk 3-rowed at the capsular base, exindusiate, usually round to linear sori and monolete, bilateral spores. The number of genera included in different systems of classification is extremely variable. Ching (1940) had recognised 45 genera, de la Sota (1973) assigned 57 genera and Pichi Sermolli (1977) recognised 63 genera in the Polypodiaceae. In South India, the family is represented by 10 genera and 23 species. The genera with the respective number of species in bracket are *Leptochilus* (4), *Microsorum* (4), *Phymatosorus* (4), *Pyrrosia* (4), *Lepisorus* (2) ; while *Colysis, Drynaria, Phlebodium, Phymatopteris* and *Pleopeltis* are represented by single species each.

Polypodiaceae Berchtold & J.C. Presl, Prirozen. Rostlin. 1 : 272. 1820.

Type genus : *Polypodium* L.

Platyceriaceae Ching, Sunyatsenia 5 : 256. 1940.

Drynariaceae Ching, Acta Phytotax. Sin. 16 : 1-37. 1978.

Rhizome short to long creeping, paleaceous; dictyostelic, often with scattered sclerenchyma strands in the ground tissue. Palea basifixed, pseudo-peltate or peltate, often clathrate, entire, toothed, ciliate or lacerate, glandular. Fronds tufted or distantly placed, mostly in two rows, simple, lobed, pinnate or pinnatifid or pedate, monomorphic or dimorphic; stipe articulate. Venation reticulate, usually with free included veins. Sori usually round, oval or linear, or occasionally acrostichoid, exindusiate, paraphysate. Sporangia short to long stalked, stalk 3-rowed at the capsular base. Spores monolete, bilateral, exine smooth or variously ornamented.

Key to the genera

1a. Humus collecting nest leaves present; abscission tissue present along the costa and between the pinna lobes............................*Drynaria*

1b. Humus collecting nest leaves absent; no abscission tissue present along the costa and between the pinna lobes......................................2

2a. Fronds bearing persistent stellate hairs..*Pyrrosia*

2b. Fronds naked, scaly or bearing non-stellate hairs.......................3

3a. Fronds distinctly dimorphic; fertile fronds narrowly elongated, with or without lamina; sori acrostichoid..........................*Leptochilus*

3b. Fronds usually monomorphic; fertile fronds not contracted; sori not as above 4
4a. Sori when young covered with umbrella-shaped paraphyses...........5
4b. Sori not covered with such paraphyses.....................................6
5a. Rhizome having numerous sclerenchyma strands in the ground tissue; abaxial surface of fronds sparsely paleaceous..............................*Lepisorus*
5b. Rhizome devoid of sclerenchyma strands in the ground tissue; abaxial surface of fronds densely paleaceous.................*Pleopeltis*
6a. Palea clathrate.......................................7
6b. Palea non-clathrate.................................8
7a. Sori sunk in cavities.................................*Phymatosorus*
7b. Sori not sunk in cavities..........................9
8a. Margins of fronds notched; included veins both excurrent and recurrent..............................*Phymatopteris*
8b. Margins of fronds not so; included veins excurrent only...............*Phlebodium*
9a. Sori continuous or more or less broken in single line between adjacent main veins.......................*Colysis*
9b. Sori small, round, scattered irregularly or in two rather regular rows between adjacent main veins.........*Microsorum*

COLYSIS Presl

Presl in 1849 established the genus *Colysis* and this was followed by Fee (1850-1852). Most of the subsequent authors (Moore, 1857; Hooker, 1864; Clarke, 1880; Christensen, 1906), however, preferred to keep its species under different genera. It was Ching (1933b, 1940), who revived the genus *Colysis* and placed 20 species under it. Since then, most authors (Copeland, 1947; Holttum, 1954; Devol & Kuo, 1975; Satija & Bir, 1985; Hennipman *et al.*, 1991) kept this as a separate genus. *Colysis* is treated here as construed by Copeland (1947).

Colysis have a superficial resemblance to *Selliguea* Bory particularly because of having similarly oblique coenosori and mostly simple, entire fronds. Hence the former is often placed under the latter. However, in the nature of palea, texture of fronds and in many other characters, the two genera are distinct. *Colysis* have clathrate palea and herbaceous to sub-coriaceous fronds as against non-clathrate palea and coriaceous fronds in *Selliguea*. Moreover, the same sorus-form in the two genera has had a different origin, *Colysis* from *Microsorum* and *Selliguea* from *Crypsinus*. (Copeland, 1947).

Colysis Presl, Epim. Bot. 146. 1849; Fee, Gen. Fil. 175. 1851; Ching, Bull. Fan Mem. Inst. Biol. 4 : 313. 1933; Copeland, Gen. Fil. 198. 1947; Holttum, Fern Fl. Malaya 2 : 159. 1954; Devol & Kuo in Li *et al.* (eds.), Fl. Taiwan 1 : 168. 1975; Hennipman *et al.* in Kubitzki (ed.), Fam. Gen. Vas. Plants 1 : 218. 1991, in part.

Type species : *Colysis hemionitidea* (Wall.) Presl

Polypodium sect. *Leptoselliguea* C. Chr., Gardens' Bull. 7 : 309. 1934.

Rhizome creeping or climbing, sometimes rather long, sparsely or densely paleaceous; dictyostelic. Palea small, ovate-lanceolate, entire or sub-entire, clathrate. Fronds simple or digitate or pinnate with adnate pinnae, monomorphic or dimorphic; stipe commonly decurrent-winged, articulate; lamina herbaceous to sub-coriaceous, glabrous. Veins anastomosing, main lateral veins prominent, reaching almost to the margin but rather irregular, with two or sometimes more rows of areoles between the adjacent main lateral veins, free veins simple, excurrent or recurrent. Sori between adjacent main lateral veins, round or elongate, continuous or more or less broken, not sunk in cavities. Sporangial stalk long, 3-rowed at the capsular base; annulus 12-14-celled. Spores monolete, bilateral, hyaline to light brown, smooth or minutely spinescent.

Gametophyte : Under laboratory conditions, the spore germinate within one month. The young prothallus is at first spathulate and finally become strap-shaped and branched (Nayar, 1962).

Chromosome numbers : n = 36; 2n = 72, 108 (cf. Fabbri, 1965).

Distribution and ecology : *Colysis*, with about 30 species is mainly Asiatic. Dixit (1984) recorded 8 species from India of which one is distributed in South India (Manickam & Irudayaraj, 1992). During the present investigation, no species of *Colysis* was found growing wild. However, two species, namely, *C. elliptica* and *C. hemionitidea* are available in the botanical gardens and nurseries of South India. The plant usually grows on the base of trees in evergreen forests or as a lithophyte along exposed stream banks (Manickam & Irudayaraj, 1992).

Colysis hemionitidea (Wall.) Presl, Epim. Bot. 147. 1849; Ching, Bull. Fan Mem. Inst. Biol. 4 : 320. 1933 & Ic. Fil. Sin. 4 : pl. 195. 1937; Nayar & Kaur, Comp. Bedd. Handb. Ferns Brit. India 87. 1974; Devol & Kuo in Li *et al.* (eds.), Fl. Taiwan 1 : 170. 1975; Satija & Bir, Polyp. Ferns India 83. 1985; Manickam & Irudayaraj, Pterid. Fl. W. Ghats 329. pl. 250. 1992.

Type : *Wallich*, Cat. no. 284 (holotype, K)

Polypodium hemionitideum Wall., Cat. no. 284. 1828, *nom. nud.* Clarke, Trans. Linn. Soc. Lond. Bot. 1 : 561. 1880; C. Chr., Ind. Fil. 532. 1906.

Selliguea hemionitidea Presl, Tent. Pterid. 216. t. 9. f. 17. 1836.

Pleopeltis hemionitidea Moore, Ind. Fil. 436. 1862; Beddome, Ferns South India 61. t. 182. 1864 & Handb. Ferns Brit. India 358. 1883.

Gymnopteris feei f. *anomala* Beddome, Ferns Brit. India 274. 1866.

(Fig. 15)

Rhizome long creeping, 5 mm in diameter, sparsely paleaceous. Palea ovate-lanceolate, 2 × 0.3 mm, clathrate, apex acuminate, margins slightly dentate, dark-brown. Fronds simple, distantly placed, 40-60 × 5-7 cm, articulated; stipe short, adaxially grooved, paleaceous at the base ; lamina broadly lanceolate, widest at the middle, entire, acuminate towards apex, gradually narrowed and long decurrent towards base, herbaceous, glabrous. Venation reticulate, midrib grooved, rounded below, main lateral veins distinct, nearly parallel, veins anastomosing and form areoles, included free veins once or twice forked. Sori in one row between adjacent main lateral veins, round or oblong, interrupted. Sporangial stalk 3-rowed at the apex; annulus 12-14—celled. Spores monolete, bilateral, exine smooth; aborted spores numerous.

Distribution and ecology : *C. hemionitidea* is distributed in India, Nepal, Bhutan, China, Formosa, Tonkin, Vietnam, Taiwan, Philippine islands and Ryukyus. It usually grows on rocks and tree trunks in wet, humid localities in the forests. Manickam and Irudayaraj (1992) described it as very rare species in South India (Fig. 17).

Notes : Because of the unstable soral conditions, which are generally interrupted into oblong or roundish shape, *C. hemionitidea* is often considered as a linking species between the genera *Colysis* and *Microsorum.* Hennipman *et al.* (1991) reported that the sori of this species has rarely become acrostichoid.

Specimen examined : TAMIL NADU : Nilgiri Dt., Ooty Bot. Gard. (Cult.), *Santhosh Nampy* 48656 (CALI).

DRYNARIA (Bory) J. Sm., *nom. cons.*

The genus *Drynaria* was established by J. Smith (1841) to include the species earlier treated under *Polypodium* subgen. *Drynaria* Bory. The genus is clearly defined and there has been no considerable confusion as to its circumscription or delimitation. Now it includes about 15 species.

Drynaria is unique among other polypodioid genera of South India in having fronds of two kinds : The sessile, broad, humus collecting nest leaves and long stalked, deeply-lobed foliage leaves. It also possesses some characters like thick, fleshy rhizome; humus collection of vegetative fronds; areoles without free included veins; sori at the junction of veins; abscission of segments from the rachis and non-articulated stipe.

The name 'Drynaria' possibly refers to its attractive sterile fronds resembling Oak leaves, a tree sacred to the Dryads (a wood nymph).

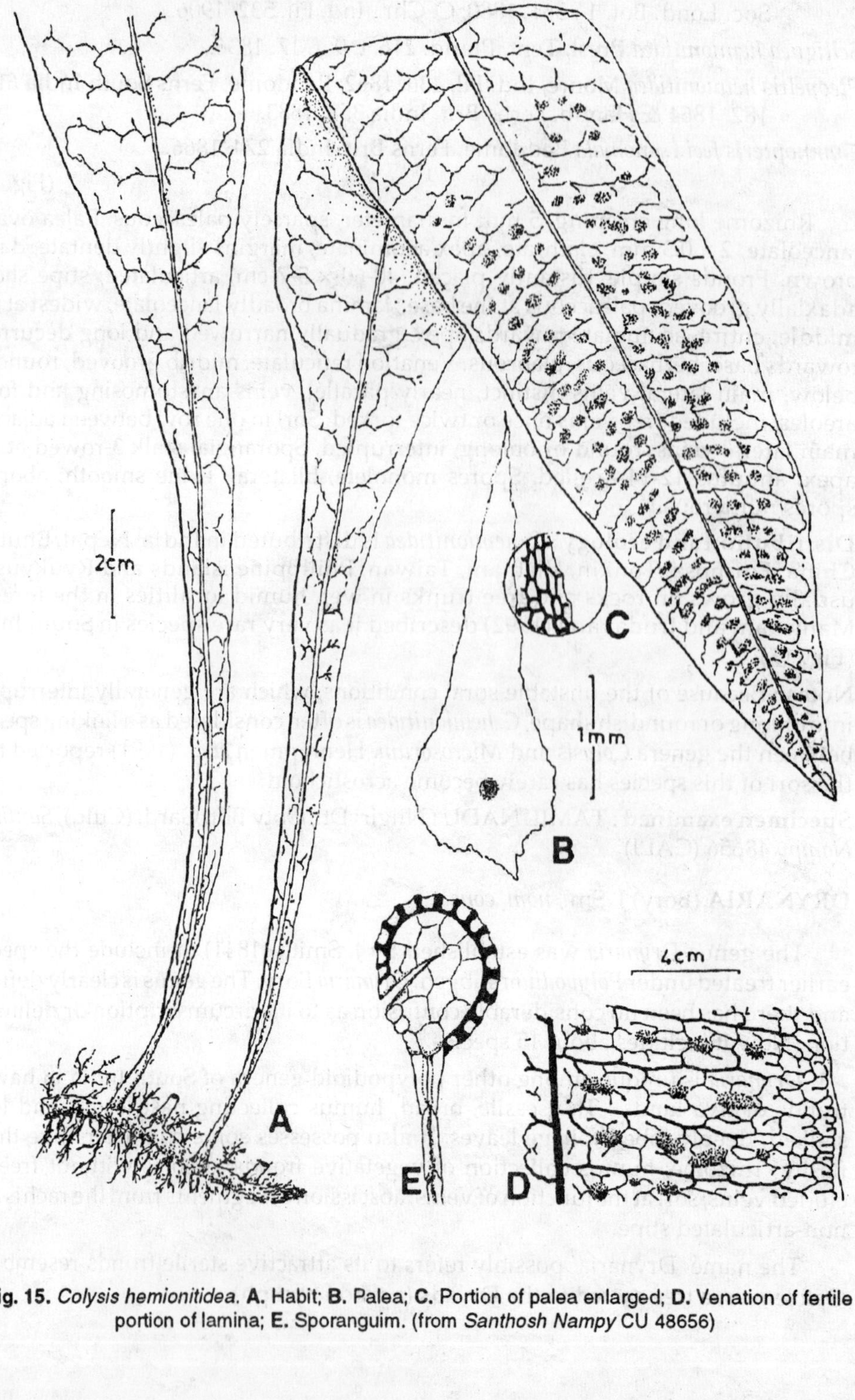

Fig. 15. *Colysis hemionitidea.* **A.** Habit; **B.** Palea; **C.** Portion of palea enlarged; **D.** Venation of fertile portion of lamina; **E.** Sporanguim. (from *Santhosh Nampy* CU 48656)

Drynaria (Bory) J. Sm., Jour. Bot. 4 : 60. 1841 ; Copeland, Gen. Fil. 203. 1947; Holttum, Fern Fl. Malaya 2 : 182. 1954; Nayar, Bull. Nat. Bot. Gard. Lucknow 56(2) : 1-16. 1961; Roos, Univ. Utrecht Publ. 255. 1985; Hennipman *et al.* in Kubitzki (ed.), Fam. Gen. Vas. Plants 1 : 212. 1991.

Polypodium subgen. *Drynaria* Bory, Ann. Sci. Nat. 5 : 463. 1825.

Polypodium sect. *Drynaria* Bory, Ann. Sci. Nat. 5 : 464. 1825.

Type species : *Drynaria quercifolia* (L.) J. Sm. (= *Polypodium quercifolium* L., *typus cons.*).

Rhizome creeping, thick, fleshy, younger parts densely paleaceous. Palea linear-lanceolate, peltate, non-clathrate, margins ciliate or dentate. Heterophyllous : nest leaves sessile, broad but short, sterile; foliage leaves long stalked, deeply lobed or pinnate, sterile or fertile, pinnules jointed to the rachis; stipe non-articulated. Venation anastomosing, veins raised, areoles without free included veins. Sori round, 2 mm in diameter, borne at the junction of the veins; paraphysis absent. Spores monolete, bilateral, exine spinulose.

Gametophyte : The mature prothallus is cordate, broader than long, usually 5-7 mm in diameter (Nayar, 1965).

Chromosome numbers : n = 36, 37 (cf. Hennipman *et al.*, 1991).

Distribution and ecology : *Drynaria* is a large epiphytic or petrophilous fern widely distributed throughout Asia to North-east Australia and Africa. Four species are reported to occur in India (Satija & Bir, 1985) of which *D. quercifolia* alone is distributed in South India.

Drynaria quercifolia (L.) J. Sm., Jour. Bot. 3 : 398. 1841; Beddome, Ferns South India 63. t. 187. 1864 & Handb. Ferns Brit. India 341. t. 191. 1883; Ching, Sunyatsenia 5 : 261. 1940; Holttum, Fern Fl. Malaya 2 : 182. 1954; Sledge, Bull. Brit. Mus. (Nat. Hist.) Bot. 2(5) : 144. 1960; Nayar, Bull. Nat. Bot. Gard. Lucknow 56 : 11. 1961; Roos, Univ. Utrecht Publ. 258. 1985; Satija & Bir, Polyp. Ferns India 88. 1985; Manickam & Irudayaraj, Pterid. Fl. W. Ghats 312. pl. 238. 1992; Nayar & Geevarghese, Fern Fl. Malabar 382. 1993.

Polypodium quercifolium L., Sp. Pl. 2 : 1087. 1753.

Type : Herb Linne (Microfiche No. 1251.12).

–Welli-panna-kelengu maravara Rheede, Hort. Malab. 12 : 25. t. 12, 13. 1693.

(Figs. 8f, 16)

Rhizome short creeping, stout, upto 3 cm in diameter, adpressed to the substratum tightly, densely clothed with paleae all over. Palea linear-lanceolate or lanceolate, *ca.* 10 × 1 mm, peltate with a short stalk, red-brown, non-clathrate, gland-tipped, gland deciduous, margins dentate-ciliate, a few scales are distributed on the lamina abaxially especially on the veins. Fronds seasonal, dimorphic, coriaceous or sub-coriaceous; nest leaves humus collecting, broad, 15 × 15 cm, cordate-ovate, simple or shallowly lobed, pseudo-peltate, sessile, closely covering the rhizome, lateral veins very prominent, glossy, green when young but soon turning grey-brown; fertile fronds 74 × 30 cm, deeply pinnatifid (juvenile leaves simple), stipitate; stipe *ca.* 17 cm long, rounded abaxially but shallowly grooved and flattened

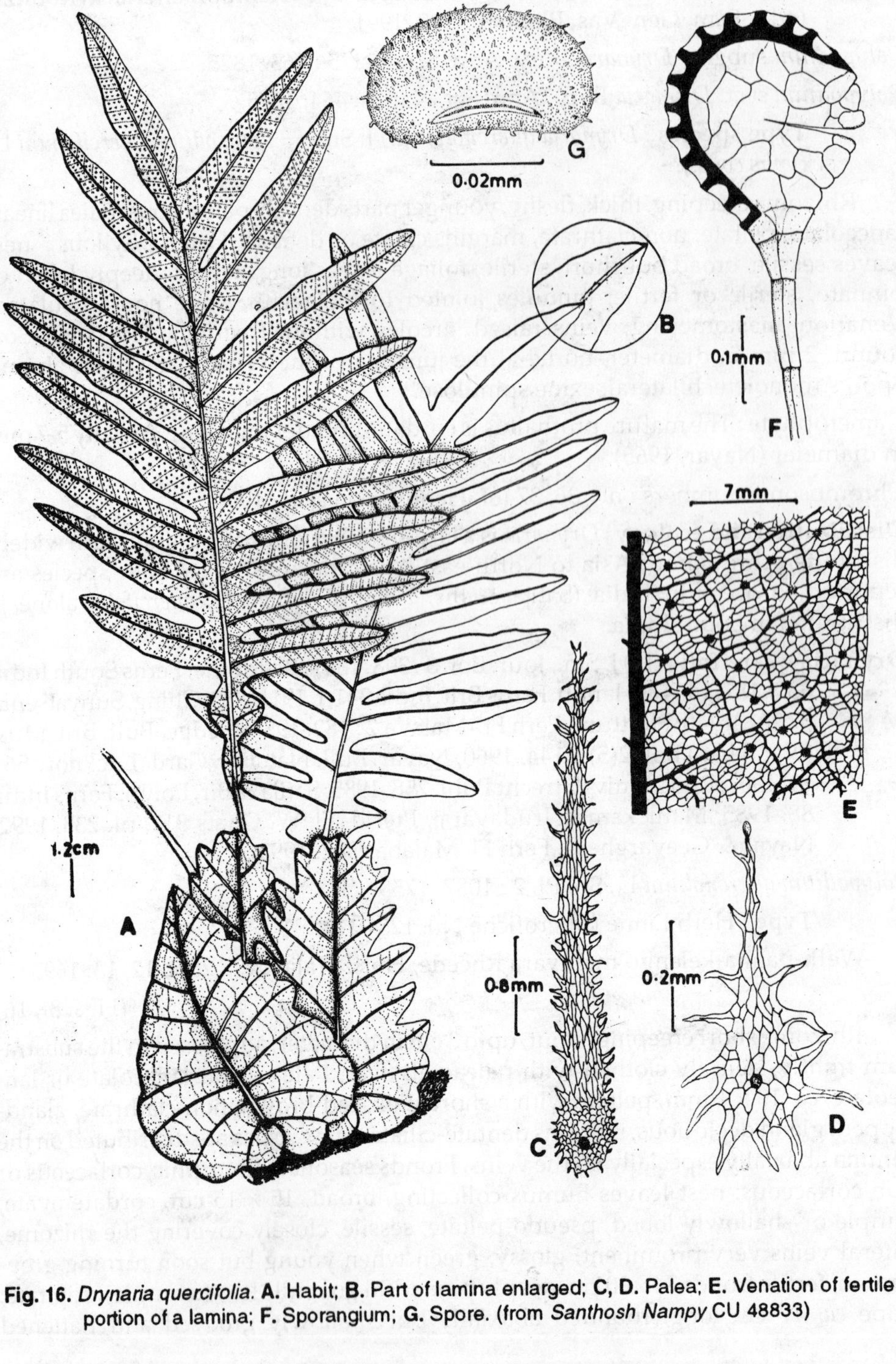

Fig. 16. *Drynaria quercifolia.* A. Habit; B. Part of lamina enlarged; C, D. Palea; E. Venation of fertile portion of a lamina; F. Sporangium; G. Spore. (from *Santhosh Nampy* CU 48833)

adaxially; leaf lobes sub-opposite, 18 × 3.5 cm, separated by rather narrow sinuses, gradually tapering from a broad base to acute or shortly acuminate apex, margins cartilaginous, slightly undulate, abscission from the rachis common when drying. Venation pinnate-reticulate, veins conspicuous and raised abaxially on the lamina, lateral veins obliquely running; primary areoles enclose a network of quadrangular secondary areoles without or rarely with a very few free veins, some of the marginal veinlets unite with the cartilaginous edge of the lamina; nectaries present close to the junction of the main veins of the segments with the midrib. Sori numerous, distinct, evenly distributed, usually two per each primary areole, round, 1.5 mm in diameter; paraphysis absent. Sporangial stalk 2.5 times longer than capsule; annulus 13-celled. Spores 60 × 36 µm, monolete, bilateral, yellowish to golden-brown, spinulose.

Distribution and ecology : *D. quercifolia* is distributed throughout India, Sri Lanka, Malesia to Polynesia and tropical Australia. It is fairly common in South India and occurs on exposed tree trunks or rocks or on old brick walls from sea level to an elevation of 900 m (Fig. 17).

Notes : *D. quercifolia* is characterised by its linear-lanceolate rhizome scales, abaxially raised venation, areoles without or with a few free included veins and almost two regular rows of sori between adjacent lateral veins. It also possesses nectaries, close to the junction of the main veins of the segments with the midrib. The plant is used in medicine. Its rhizome is bitter and astringent; fronds are used in hectic fever, dyspepsia, cough and phthesis (Nayar, 1959; Dixit & Vohra, 1984).

Specimens examined : KARNATAKA : Chikmagalur Dt., Sringeri, *Santhosh Nampy* 48833 (CALI). KERALA : Idukki Dt., Edpalayam, *Sharma* 41664 (MH); Kaliyar, *Santhosh Nampy* 48658 (CALI); way to Mangaladevi temple, *Vivekananthan* 50541 (MH). Kannur Dt., Karimbam, *Ansari* 67841 (MH). Kottayam Dt., Vandiperiyar, *Vivekananthan* 21338 (MH). Kozhikode Dt., Medical College campus, *Santhosh Nampy* 48657 (CALI); Kutiyadi, *Naithani* 24188 (MH). Malappuram Dt., Thiruvangad, *Ravindran* 18133 (CALI). Palakkad Dt., Dhoni hills, *Geevarghese* 4144 (CALI); Kaikatty to Pothundi, *Vajravelu* 44786; Olavakode, *Vajravelu* 19143 (MH). Thiruvananthapuram Dt., way to Agasthiamala, *Santhosh Nampy* 48616; Athirumala, *Santhosh Nampy* 48615 (CALI); Kottur R.F., *Joseph* 44455 (MH); Ponmudi, *Madhusoodanan* 43855 (CALI). Wynad Dt., Chandanathode, *Ellis* 25207, *Ramachandran* 57600 (MH). TAMIL NADU : Coimbatore Dt., Amarawatisagaram, *Deb* 30144 (MH). Kanyakumari Dt., Veerapulli R.F., *Henry* 48238 (MH). Kattabomman Dt., Courtallam, *Subramanyam* 2899 (MH). Madurai Dt., Kumbukere, *Sebastiane* 12630; near Suruli falls, *Subramanyam* 8190 (MH). Nilgiri Dt., Benne forest, *Subramanyam* 10466; Gudalur, *Vajravelu* 41752; Mudumalai R.F., *Sebastiane* 7368; northern Hay R.F., *Sharma* 33507 (MH).

LEPISORUS (J. Sm.) Ching

Smith (1846) treated *Lepisorus* as a section under the genus *Drynaria* (Bory) J. Sm. Later, Christensen (1906) transferred it from *Drynaria* and placed again as a section (sect. *Lepisorus*) under the subgen. *Pleopeltis* of the genus *Polypodium* L. Ching (1933a) raised it to the rank of a genus, regarding it as phyletically distinct from *Pleopeltis*, which according to him is purely an American genus. Since then, the circumscription of the genus *Lepisorus* has been a matter of great dispute. Christensen

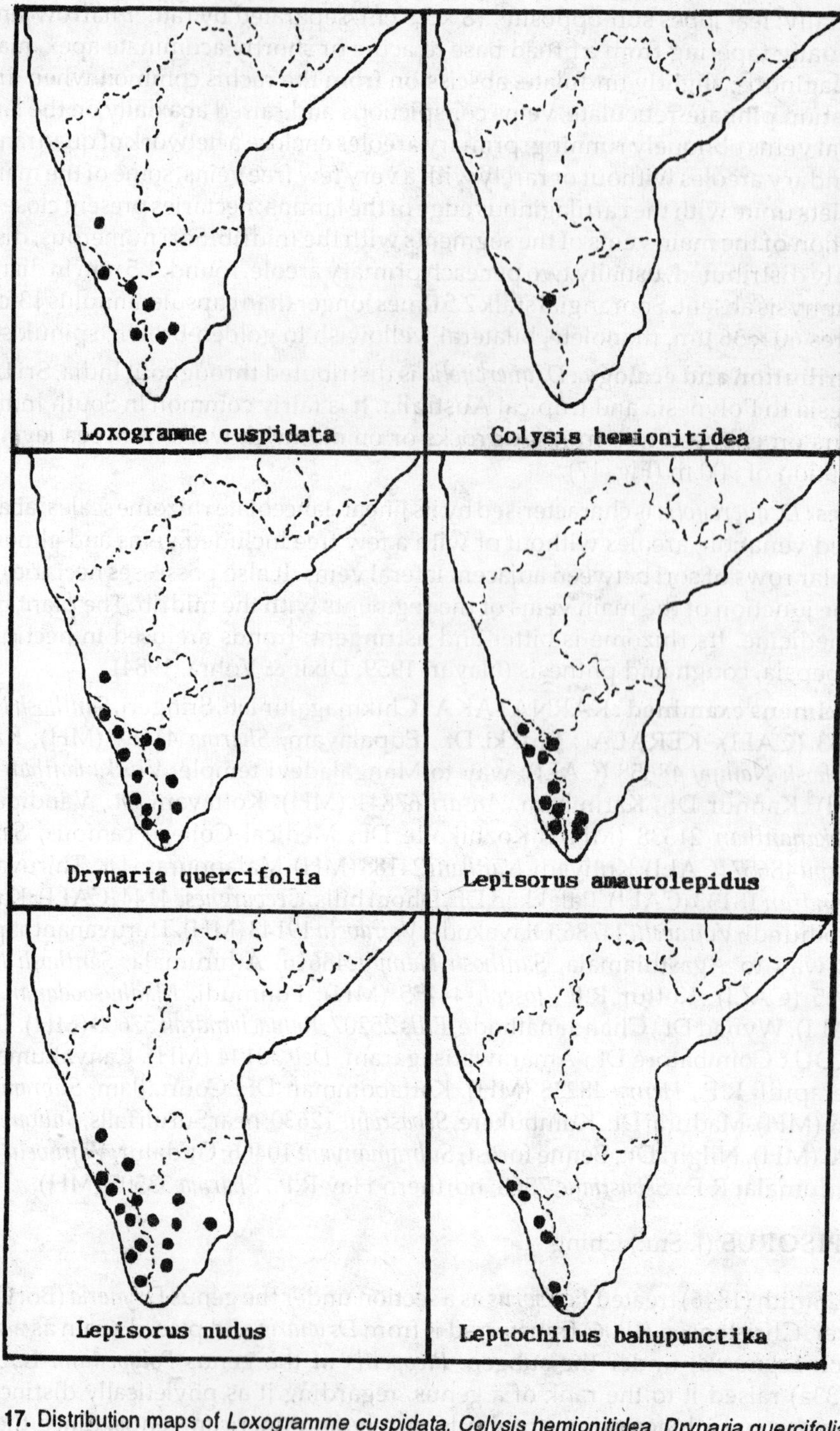

Fig. 17. Distribution maps of *Loxogramme cuspidata*, *Colysis hemionitidea*, *Drynaria quercifolia*, species of *Lepisorus* and *Leptochilus bahupunctika* in South India

(1934) reduced it to the subgeneric status under *Polypodium*. However, Copeland (1947), Sledge (1960) and Pichi Sermolli (1977) treated *Lepisorus* as conspecific with *Pleopeltis*.

Ching (1933a, 1940) and later Holttum (1954), Bir and Trikha (1969, 1974), de la Sota (1973), Tryon and Tryon (1982), Satija and Bir (1985), Hennipman *et al*. (1991), and Manickam and Irudayaraj (1992) have recognised *Lepisorus* as distinct from *Pleopeltis*. Zink (1988) elaborately discussed the reasons for keeping these two genera separate. *Lepisorus* can be easily distinguished from *Pleopeltis* in having numerous sclerenchyma strands in the rhizome, totally clathrate palea, abaxial surface of fronds sparsely paleate as against sclerenchyma strands absent in the rhizome, margins of palea not clathrate and the abaxial surface of fronds densely paleaceous in *Pleopeltis*. Apart from these, they differ in their geographical distribution also, *Lepisorus* is Asiatic while *Pleopeltis* is Neotropical in distribution.

Lepisorus is a taxonomically difficult genus because many of its species are feebly distinguished morphologically. Species delimitation is principally based on microcharacters such as scale morphology and spore characters.

The name 'Lepisorus' is possibly derived from the Greek, 'lepis' meaning a scale, referring to its characteristic scale-like paraphyses covering the immature sorus.

Lepisorus (J. Sm.) Ching, Bull. Fan Mem. Inst. Biol. 4 : 47. 1933; C. Chr., Verdoorn's Manual Pterid. 547. 1938; Holttum, Fern Fl. Malaya 2 : 151. 1954; Bir & Trikha, Bull. Bot. Surv. India 11 : 260. 1969; Satija & Bir, Polyp. Ferns India 17. 1985; Hennipman *et al*. in Kubitzki (ed.), Fam. Gen. Vas. Plants 217. 1991; Manickam & Irudayaraj, Pterid. Fl. W. Ghats 322. 1992; Nayar & Geevarghese, Fern Fl. Malabar 387. 1993.

Drynaria sect. *Lepisorus* J. Sm., Comp. Bot. Mag. 72 : 13. 1846.

Type species : *Lepisorus thunbergianus* (Kaulf.) Ching

Polypodium subgen. *Pleopeltis* sect. *Lepisorus* C. Chr., Ind. Fil. 498. 1906.

Rhizome creeping, slender, paleaceous; dictyostelic, isolated sclerenchyma strands distributed in the ground tissue. Palea clathrate, central part generally opaque. Fronds in two dorsal rows on the rhizome; lamina simple, both surfaces naked or with a few scales; stipe short, articulated. Veins hidden, secondary veins catadromous, areoles irregular, included and a few excluded free veins present. Sori in one row on either side of the midrib, round or elliptic, large, protected in young stage by characteristic umbrella-shaped clathrate paraphyses. Spores monolete, bilateral, smooth or tuberculate.

Gametophyte : Prothallial morphology of a few species of *Lepisorus* is known (Nayar, 1961d, 1962). The adult prothallus is cordate with a deep apical notch and spreadout lateral wings.

Chromosome numbers : x = 22, 23, 25, 26, 35, 36, 37, 47 (cf. Fabbri, 1965).

Distribution and ecology : *Lepisorus* with about 25 species is widely distributed in the tropical and sub-tropical regions of the Old World. Most of its species grow either as a lithophyte or epiphyte in mossy, humid areas of medium to higher elevations. Bir and Trikha (1969) made detailed studies on this genus in India and listed 20 species. According to them, three species were present in South India, *viz.*, *L. amaurolepidus* (Sledge) Bir & Trikha, *L. excavatus* (Bory) Ching and *L. nudus* (Hook.) Ching. However, no specimens of *L. excavatus* from South India were available at CAL or MH. It has not so far been reported from South India, although Bir and Trikha (1969) considered an extension of its distribution in the southern parts of India. In the present treatment two species of *Lepisorus, viz., L. amaurolepidus* and *L. nudus* are described.

Key to the species

1a. Rhizome scales bicolorous, margins dentate.......................*L. amaurolepidus*

1b. Rhizome scales concolorous, margins entire.......................*L. nudus*

Lepisorus amaurolepidus (Sledge) Bir & Trikha in Bir & Vasudeva, Jour. Bombay Nat. Hist. Soc. 68 : 192. 1971; Bir & Trikha, Bull. Bot. Surv. India 11 : 266. 1972 (1969 issue) & Amer. Fern Jour. 64(2) : 60. 1974; Satija & Bir, Polyp. Ferns India 19. 1985; Manickam & Irudayaraj, Pterid. Fl. W. Ghats 334. 1992; Nayar & Geevarghese, Fern Fl. Malabar 387. pl. 253. 1993.

Pleopeltis amaurolepida Sledge, Bull. Brit. Mus. (Nat. Hist.) Bot. 2(5) : 136. 1960; Manickam, Fern Fl. Palni Hills 157. 1986.

Type : Sri Lanka, *Sledge* 999 (holotype, BM).

Polypodium gladiatum Wall., Cat. no. 279. 1829, *nom. nud.*, non Kunze, 1834.

Lepisorus amaurolepidus var. *longifolius* Bir & Trikha, Amer. Fern Jour. 64(2) : 63. 1974.

(Figs. 8 e, 18)

Rhizome short creeping, 2 mm in diameter, densely paleaceous; dictyostelic, sclerenchyma strands numerous in the ground tissue. Palea ovate-lanceolate, 2 × 1 mm, peltate, clathrate, dark-brown, bicolorous, walls of central cells highly thickened and their lumen completely filled with depositions, marginal cells hyaline, gland tipped, dentate. Fronds monomorphous, closely placed, 1-6 mm apart, in two dorsal rows on the rhizome, shortly stipitate; stipe below 0.5 cm, naked, articulated; lamina simple, entire, lanceolate, *ca.* 8 × 1 cm, coriaceous or herbaceous, a few scales similar to rhizome scales are present abaxially, apex acute. Venation indistinct externally, anastomosing, secondary veins catadromous, areoles irregular, excluded and included free veins are with swollen tips. Sori median, in one row on either side of the midrib, distinct, 2.5 mm in diameter, round; paraphyses umbrella-shaped, peltate, long-stalked, completely cover the sori during juvenile stage, clathrate, dark-brown, walls of central cells highly thickened, marginal cells paler, glandular. Annulus 12-celled. Spores 60 × 47 µm, elliptic-oblong, monolete, bilateral, pale yellow, tuberculate.

Distribution and ecology : *L. amaurolepidus* is distributed in India and Sri Lanka. It occurs either as a lithophyte on wet rocks on the forest floor or as an epiphyte in humid situations at an altitude above 800 m (Fig. 17).

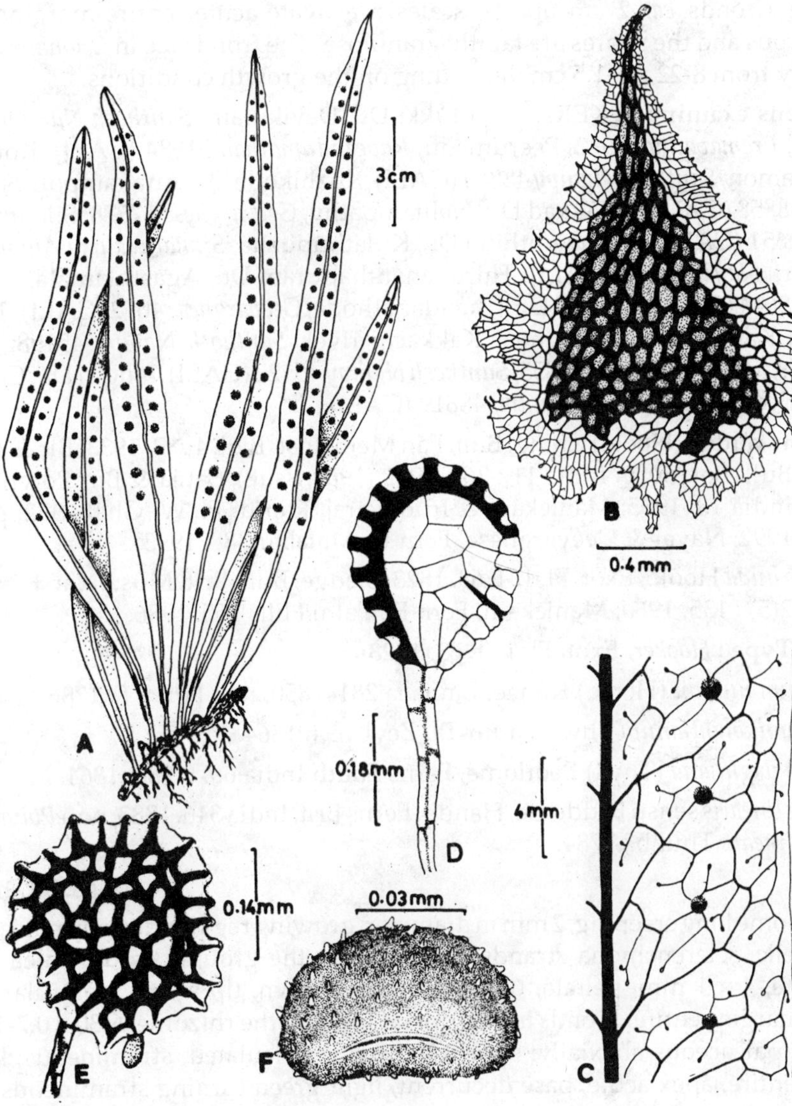

Fig. 18. *Lepisorus amaurolepidus.* **A.** Habit; **B.** Palea; **C.** Venation of fertile portion of a lamina; **D.** Sporangium; **E.** Paraphysis; **F.** Spore. (from *Santhosh Nampy* CU 51981)

Notes : *L. amaurolepidus* is closely allied to *L. nudus* but can readily be distinguished by its ovate-orbicular, dentate-margined, bicolorous scales, short creeping rhizome (fronds 1-6 mm apart) and tuberculate spores. In *L. nudus*, the rhizome is long creeping (fronds *ca.* 2 cm apart), scales are ovate-acute, entire-margined and concolorous and the spores are faintly granulose. The frond size in *L. amaurolepidus* may vary from 8-22 × 1-1.5 cm depending on the growth conditions.

Specimens examined : KERALA : Idukki Dt., Devikulam, *Santhosh Nampy* 51981; Munnar, *Premavathy* 51982; Peerumedu, *Raveendranathan* 51984 (CALI). Kottayam Dt., Vagamon, *Santhosh Nampy* 48864 (CALI). Kozhikode Dt., Anakampoil, *Santhosh Nampy* 48858 (CALI). Palakkad Dt., Nelliampathy, *Geevarghese* 17878; Silent Valley, *Nayar* 10651 (CALI). Pathanamthitta Dt., Kulathupuzha, *Sumathykutty Amma* 5969; Sabarigiri, *Nayar* 10138 (CALI). Thiruvananthapuram Dt., Agasthiamala, *Santhosh Nampy* 48614 (CALI). Wynad Dt., Chandanathode, *Geevarghese* 4092 (CALI). TAMIL NADU : Kattabomman Dt., near Kakkachi river, *Santhosh Nampy* 45198; Upper Kothayar, *Madhusoodanan* 44855, *Santhosh Nampy* 45122 (CALI). Nilgiri Dt., Coonoor Sims park (wild), *Santhosh Nampy* 48819 (CALI).

Lepisorus nudus (Hook.) Ching, Bull. Fan Mem. Inst. Biol. 4 : 83. 1933; Bir & Trikha, Bull. Bot. Surv. India 11 : 265. 1972 (1969 issue); Satija & Bir, Polyp. Ferns India 18. 1985; Manickam & Irudayaraj, Pterid. Fl. W. Ghats 332. pl. 252. 1992; Nayar & Geevarghese, Fern Fl. Malabar 388. 1993.

Pleopeltis nuda Hook., Exot. Fl. 1 : t. 63. 1823; Sledge, Bull. Brit. Mus. (Nat. Hist.) Bot. 2(5) : 135. 1960; Manickam, Fern Fl. Palni Hills 155. 1986.

Type : *Hooker,* Exot. Fl. 1 : t. 63. 1823.

Polypodium nudum (Hook.) Kunze, Linnaea 281. 1850, non Forst. f., 1786.

Polypodium wightianum Thw., Enum. Pl. Zeyl. 394. 1864.

Pleopeltis wightiana (Thw.) Beddome, Ferns South India 60. t. 180. 1864.

Pleopeltis linearis sensu Beddome, Handb. Ferns Brit. India 346. 1883, non *Polypodium lineare* Thunb., 1784.

(Figs. 8d, 14e, 19)

Rhizome long creeping, 2 mm in diameter, growing regions densely paleaceous; dictyostelic, sclerenchyma strands numerous in the ground tissue. Palea ovate-lanceolate, 4 × 1 mm, peltate, uniformly pale brown, tip acute, glandular when young, margins entire. Fronds in two dorsal rows on the rhizome, 6-30 × 0.7-1.6 cm, sparsely paleaceous abaxially; stipe 3 × 0.1 cm, articulated, stramineous; lamina simple, entire, apex acute, base decurrent, light green turning stramineous when dry, coriaceous or sub-coriaceous. Veins indistinct externally, included and excluded free veins with swollen tips. Sori circular or ellipsoidal, 3 mm in diameter, one row on either side of the midrib, near to the costa than the margins, one per each costal areole; paraphyses umbrella-shaped, peltate, long-stalked, clathrate, glandular. Annulus 14-celled, sporangial stalk 3-rowed at the capsular base. Spores 78 × 50 µm monolete, bilateral, yellowish or hyaline, faintly granulose.

Distribution and ecology : *L. nudus* is distributed in North and South India, Sri Lanka, China and Malay Peninsula. It frequently occurs on mossy tree trunks and

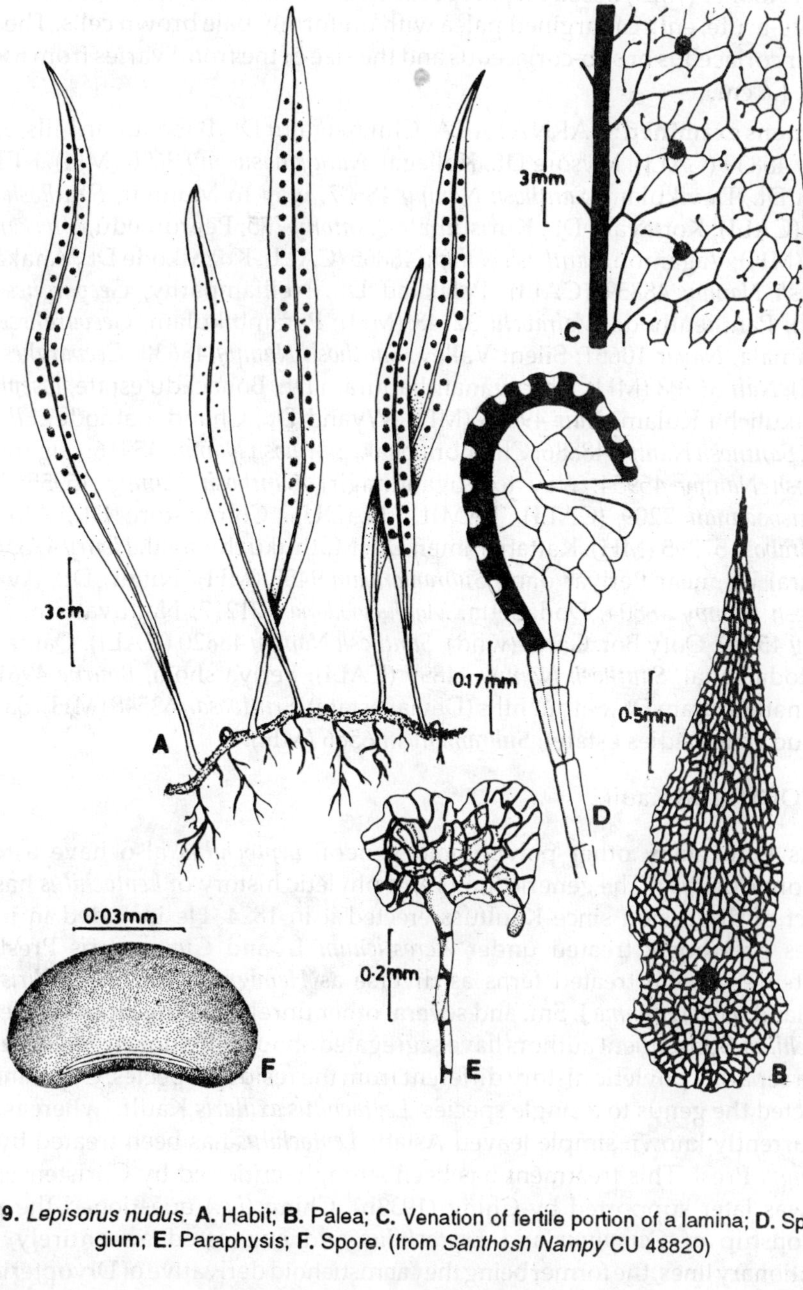

Fig. 19. *Lepisorus nudus.* **A.** Habit; **B.** Palea; **C.** Venation of fertile portion of a lamina; **D.** Sporangium; **E.** Paraphysis; **F.** Spore. (from *Santhosh Nampy* CU 48820)

wet boulders in the mountains (above 700 m) of South India (Fig. 17). It is also found growing as an epiphyte on tea bushes or on shola trees.

Notes : Hooker (1823) based Nepal specimen for his *L. nudus* which is characterised by ovate-acute, entire margined palea with uniformly pale brown cells. The lamina is either coriaceous or sub-coriaceous and the size of the frond varies from 6 × 0.7 cm to 30 × 1.6 cm.

Specimens examined : KARNATAKA : Chikmagalur Dt., Bababudan hills, *Santhosh Nampy* 48839 (CALI). Mysore Dt., Kollegal, *Narayanaswamy* 9690 (MH). KERALA : Idukki Dt., Devikulam, *Santhosh Nampy* 48607; way to Munnar, *Santhosh Nampy* 48608 (CALI). Kottayam Dt., Kurisumalai, *Antony* 775; Peerumedu, *Vivekananthan* 21420 (MH); Vagamon, *Santhosh Nampy* 48865 (CALI). Kozhikode Dt., Anakampoil, *Santhosh Nampy* 48859 (CALI). Palakkad Dt., Nelliampathy, *Geevarghese* 17885 (CALI); Panthenthode, *Vajravelu* 32148 (MH); Parambikulam, *Geevarghese* 17880; Punnamala, *Nayar* 10651; Silent Valley, *Santhosh Nampy* 48630, *Geevarghese* 17961 (CALI), *Nair* 36708 (MH). Thiruvananthapuram Dt., Bonacadu estate, *Joseph* 44490; Seethakulicha Kulam, *Nair* 49876 (MH). Wyand Dt., Chandanathode, *Ellis* 25251 (MH), *Santhosh Nampy* 48886; Chembra peak, *Santhosh Nampy* 45916; Sugandhagiri, *Santhosh Nampy* 45954; way to Sugandhagiri, *Santhosh Nampy* 45955; Vythiri, *Madhusoodanan* 7209 (CALI). TAMIL NADU : Coimbatore Dt., Akkamalai, *Chandrabose* 57295 (MH). Kattabomman Dt., Muthukuzhivayal, *Henry* 47554 (MH). Madurai Dt., near Periyar dam, *Subramanyam* 9457 (MH). Nilgiri Dt., Avalanchi, *Santhosh Nampy* 48804; Dodabetta, *Madhusoodanan* 21217; Naduvattam, *Santhosh Nampy* 45180; Ooty Bot. Gard. (wild), *Santhosh Nampy* 48820 (CALI). Qaide Milleth Dt., Kodaikanal, *Santhosh Nampy* 44899 (CALI); Periya shola, *Bourne* 4961 (MH). Ramanathapuram Dt., Sethur hills (Deviar estate), *Srinivasan* 63548 (MH). Salem Dt., Yercaud (Babmadies estate), *Subramanyan* 6586 (MH).

LEPTOCHILUS Kaulf.

As with many other polypodioid genera, *Leptochilus* also have a tortuous taxonomic history. The generic limits and phyletic history of *Leptochilus* has been a subject of discussion since Kaulfuss erected it in 1824. He included in it certain species previously treated under *Acrostichum* L. and *Gymnopteris* Presl. Later, Christensen (1906) treated ferns as diverse as *Hemigramma* Christ, *Christiopteris* Copeland, *Lomagramma* J. Sm. and several other unrelated American species under *Leptochilus*. Subsequent authors have segregated about half a dozen genera, claiming each a separate phyletic history different from the residual species. Copeland (1928) restricted the genus to a single species, *Leptochilus axillaris* Kaulf., whereas most of the currently known simple leaved Asiatic *Leptochilus* has been treated by him as *Campium* Presl. This treatment has been strongly criticised by Christensen (1931) and was later supported by Ching (1933b). Ching (*l.c.*) questioned the phyletic relationship of *Campium* and *Leptochilus* and suggested an entirely distinct evolutionary lines, the former being the "acrostichoid derivative of Dryopteridaceae" and the latter of a polypodioid origin. He considered *Leptochilus* very close to the genus *Colysis* Presl from which it is difficult to distinguish except for its markedly dimorphic leaves and linear continuous sori, paral'-' to the midrib. He also considered the coenosoral condition as shown by *Leptochilus* may be resulted both from the specialization of functions and from the contraction of the fertile blade, thereby the normal oblique gymnogrammoid sori of *Colysis* may meet end by end in most cases.

Leptochilus as recognised by Christensen (1931) and Ching (1933b) is clearly defined and included about 10 species from tropical Asia. Ching (1933b) divided the genus into two sections, namely, sect. *Myuropteris* (C. Chr.) Ching and sect. *Euleptochilus* Ching based on the shape of sterile leaves. However, this sectional classification is not applicable for the Indian species of the genus. Sledge (1956) while revising the Sri Lankan species of *Leptochilus,* has rightly commented upon a similar situation "other characters used have been size and shape of fronds, presence or absence of stipe and texture. These characters are to different degrees all prone to vary according to environmental circumstances even in ferns which are not intrinsically variable and they would only be expected to furnish reliable criteria for taxonomic purposes if they showed either a high degree of individual constancy or consistent correlations". Christensen's and Ching's interpretations of *Leptochilus* was adopted by Holttum (1954) and Sledge (1956) which is followed here for the present treatment.

Nayar *et al.* (1985) established the genus *Nistarika* to include a single species *N. bahupunctika.* They distinguished it from *Leptochilus* by its four rows of fronds restricted to the dorsal surface of the rhizome and lamina-less fertile fronds bearing sporangia directly on the midrib in two lateral linear sori. The present study shows that the two genera are connected with a large number of intermediary forms showing many characters in common making a clear cut generic distinction not always possible. This is in corroboration with the opinion of Hennipman *et al.* (1991) that "...*Nistarika* does not constitute a distinct genus". Hence it is treated here under the synonymy of *Leptochilus.*

The name 'Leptochilus' is derived from the Greek, 'leptos', slender, and 'cheilos', a lip; possibly referring to its long slender fertile fronds usually with very narrow lamina almost entirely covered with sporangia.

Leptochilus Kaulf., Enum. Fil. 147. 1824; C. Chr., Ind. Fil. 284. 1906, in part; Copeland, Philipp. Jour. Sci. 37(4) : 338. 1928 & Univ. Calif. Publ. Bot. 16 : 114. 1929; C. Chr., Contr. U.S. Nat. Herb. 26 : 325. 1931 ; Ching, Bull. Fan Mem. Inst. Biol. 4 : 336. 1933; Copeland, Gen. Fil. 197. 1947; Holttum, Fern Fl. Malaya 2 : 163. 1954; Sledge, Ann. & Mag. Nat. Hist. Ser. 12(9) : 865. 1956 & Bull. Brit. Mus. (Nat. Hist.) Bot. 2(5) : 141. 1960; Nayar, Amer. Fern Jour. 50(4) : 301. 1963; Devol & Kuo in Li *et al.* (eds.), Fl. Taiwan 1 : 188. 1975; Satija & Bir, Polyp. Ferns India 80. 1985; Hennipman *et al.* in Kubitzki (ed.), Fam. Gen. Vas. Plants 1 : 219. 1991; Manickam & Irudayaraj, Pterid. Fl. W. Ghats 306. 1992; Nayar & Geevarghese, Fern Fl. Malabar 389. 1993.

Type species : *Leptochilus axillaris* (Cav.) Kaulf.

Acrostichum Cav., Ann. Hist. Nat. 1 : 101. 1799; Hook., Sp. Fil. 5 : 276. 1864.

Campium Presl, Tent. Pterid. 238. 1836, in part, excl. type.

Gymnopteris Presl, Tent. Pterid. 244. 1836; Beddome, Handb. Ferns Brit. India 429. 1883, in part.

Dendroglossum Presl, Epim. Bot. 149. 1849; Copeland Gen. Fil. 199. 1947, in part.

Anapausia Presl, Epim. Bot. 186. 1849, in part.

Gymnogramme Cesati, Atti Acad. Napoli 78 : 29. 1867.

Christiopteris Christ, Jour. Bot. France 21 : 273. 1908.

Drymoglossum Christ, Lecomte Not. Syst. 1 : 375. 1911.

Myuropteris C. Chr., Dansk Bot. Archiv. 6 : 73. 1929.

Paraleptochilus Copeland, Gen. Fil. 198. 1947; Nayar, Amer. Fern Jour. 50(4) : 301. 1963; Satija & Bir, Polyp. Ferns India 81. 1985.

Nistarika Nayar, Madhusoodanan & Molly, Fern Gaz. 13(1) : 33. 1985.

Rhizome creeping or climbing, dictyostelic, with or without sclerenchyma strands in the ground tissue. Palea dark-brown, clathrate, deciduous. Fronds dimorphic, in 2 or 4 dorsal rows on the rhizome, articulated; sterile fronds ovate to lanceolate, sessile or stipitate, simple; fertile fronds filiform, often longer than the sterile one, long stipitate, with or without lamina; paraphysate or non-paraphysate. Sporangia acrostichoid, either on the fertile lamina or on the groove of the midrib. Spores monolete, bilateral, densely spinose.

Gametophyte : The adult prothallus in *Leptochilus* is ribbon-like, branched and perennial (Nayar, 1963a; Nayar *et al.*, 1985).

Chromosome numbers : n = 36 (Manickam & Irudayaraj, 1988); n = 72 (cf. Chiarugi, 1960); 2n = 72 (cf. Fabbri, 1965).

Distribution and ecology : *Leptochilus,* a genus of about 12 species, is confined to tropical Asia and Malesia. Four species occur in South India, of which one, *L. bahupunctika* is endemic to the study area. It usually grows on fully or partially shaded rocky beds in mountain streams or as low epiphyte in evergreen forests.

Key to the species

1a. Rhizome devoid of sclerenchyma strands in the ground tissue........*L. thwaitesianus*

1b. Rhizome having slender sclerenchyma strands scattered in the ground tissue.................................2

2a. Fronds produced in four dorsal rows on the rhizome; fertile fronds clustered at the apex........*L. bahupunctika*

2b. Fronds produced in two dorsal rows on the rhizome; fertile fronds not as above..............3

3a. Fronds sessile, coriaceous, deep bluish-green when fresh...............*L. metallicus*

3b. Fronds stipitate, herbaceous,
uniformly green when fresh............*L. decurrens*

Leptochilus bahupunctika (Nayar, Madhusoodanan & Molly) Nampy *comb. nov.*

Nistarika bahupunctika Nayar, Madhusoodanan & Molly, Fern Gaz. 13(1) : 33-42. 1985; Nayar & Geevarghese, Fern Fl. Malabar 404. 1993.

Type : South India, *Geevarghese & Nayar* 29263 (holotype, CAL; isotype, CALI).

–Valli-varakody-maravara Rheede, Hort. Malab. 12 : 59. t. 30. 1693.

(Figs. 20, 25a)

Rhizome climbing erect, 4 mm in diameter, densely paleaceous at the apex and less so in older regions; dictyostelic, sclerenchyma strands present in the ground tissue. Palea lanceolate, 1.7×0.6 mm, clathrate, peltate, gland-tipped, base variously lobed; lobes glandular. Fronds dimorphic, in more than two dorsal rows; stipe *ca.* 3.7 cm in sterile fronds, *ca.* 22 cm in fertile fronds, grooved abaxially, articulated; sterile fronds : lanceolate, 50 × 4 cm, glabrous, gradually decurrent, apex acute; lamina simple, occasionally lobed, margins slightly recurved on drying to form cartilaginous edge, herbaceous but become stiff on drying; fertile fronds : seasonal, clustered, filiform, *ca.* 60 × 2 mm, lamina-less; paraphyses uniseriate, gland-tipped. Venation reticulate, lateral veins zigzag, areoles irregular, included and excluded free veins simple or once-forked with swollen tips. Sori acrostichoid, borne directly on the lateral groove of the midrib. Sporangial stalk slender; annulus 12-14-celled. Spores 56 × 37 μm, monolete, bilateral, exine densely spinose.

Distribution and ecology : *L. bahupunctika* is endemic to South India (Fig. 17). It has almost wider distribution throughout the evergreen forests on the Western Ghats of Kerala usually at an altitude of 200-700 m. The plant occurs in semi-shaded places, climbing vertically upon the base of shrubs and trees. The lateral roots clasp the rhizome tightly to the trunks, while a few roots grow downwards and ramify in the soil.

Notes : This species was first described by Nayar *et al.* (1985) under *Nistarika*. However, the characters such as lamina-less fertile frond and vertical climbing habit are not peculiar to this species alone as given by Nayar *et al.* but are also seen in other species of *Leptochilus*. Hence this species is transferred to the genus *Leptochilus*. *L. bahupunctika* is characterised by the fertile fronds clustered at the apex, fronds produced in more than two rows on the rhizome and the sterile fronds narrowly-lanceolate.

Specimens examined : KERALA : Malappuram Dt., Calicut Univ. Bot. Gard. (Cult.), *Madhusoodanan* 29662 (CALI). Palakkad Dt., way to Kummattanthode, *Santhosh Nampy* 51993; Kummattanthode, *Nayar* 10027, *Santhosh Nampy* 45986; Panthenthode, *Santhosh Nampy* 51996, *Nayar & Geevarghese* 29263; Parambikulam, *Geevarghese* 17881; Silent Valley (submergible area), *Nayar & Madhusoodanan* 21379 (CALI). Pathanamthitta Dt., Anathode, *Nayar* 10113; Kallipara, *Nayar* 10154; Ponnambalamedu, *Nayar* 10140 (CALI). Thiruvananthapuram Dt., way to Agasthiamala, *Santhosh Nampy & Madhusoodanan* 48612; Athirumala, *Santhosh Nampy* 51997 (CALI). Wynad Dt., Chandanathode, *Santhosh Nampy* 51998; Periya, *Santhosh Nampy* 45967, *Madhusoodanan* 29747 (CALI).

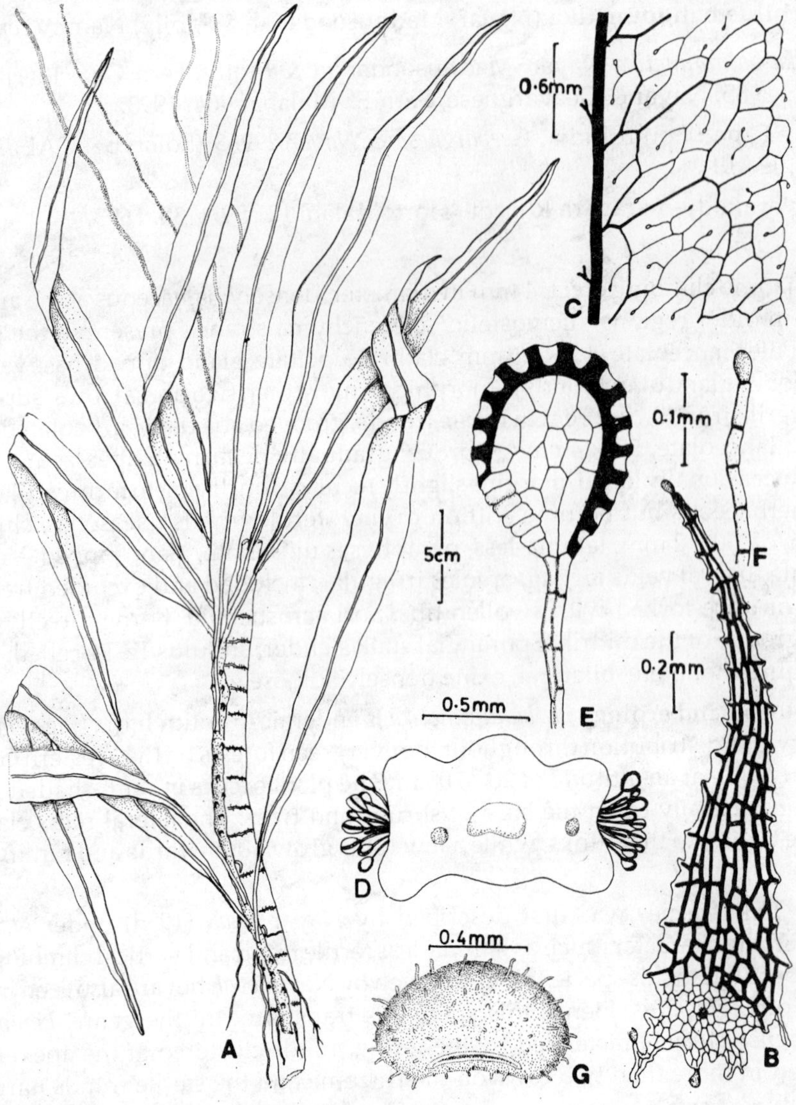

Fig. 20. *Leptochilus bahupunctika.* **A.** Habit; **B.** Palea ; **C.** Venation; **D.** Fertile frond c.s. showing attachment of sporangia; **E.** Sporangium; **F.** Paraphysis; **G.** Spore. (from *Santhosh Nampy* CU 51996)

Leptochilus decurrens Blume, Enum. Pl. Jav. 2 : 206. 1828; C. Chr., Ind. Fil. 385. 1906; Holttum, Fern Fl. Malaya 2 : 164. 1954; Sledge, Ann. & Mag. Nat. Hist. Ser. 12(9) : 867. 1956 & Bull. Brit. Mus. (Nat. Hist.) Bot. 2(5) : 140. 1960; Nayar & Kaur, Comp. Bedd. Handb. Ferns Brit. India 100. 1974; Devol & Kuo in Li *et al.* (eds.), Fl. Taiwan 1 : 189. 1975; Manickam, Fern Fl. Palni Hills 146. 1986; Chandra & Kaur, Nomen. Guide Bedd. Ferns South India & Ferns Brit. India 69. 1987; Manickam & Irudayaraj, Pterid. Fl. W. Ghats 307. pl. 234. 1992; Nampy & Madhusoodanan, Indian Fern Jour. 9 : 211. 1992; Nayar & Geevarghese, Fern Fl. Malabar 389. 1993.

Type : Java, *Blume s.n.* (holotype, Herb. Lugd. Bat.)

Leptochilus lanceolatus Fee, Acrost. 87. t. 47. f. 1. 1845; C. Chr., Ind. Fil. 386. 1906; Ching, Bull. Fan Mem. Inst. Biol. 4 : 349. 1933; Satija & Bir, Polyp. Ferns India 81. 1985; Manickam, Fern Fl. Palni Hills 147. 1986, in part.

Anapausia decurrens (Bl.) Presl, Epim. Bot. 186. 1849; Copeland, Amer. Fern Jour. 11 : 18. 1950.

Gymnopteris feei Moore, Ind. Fil. 29. 1857; Beddome, Ferns South India 15. t. 48. 1864.

Gymnopteris decurrens (Bl.) Hook., Garden Ferns t. 6. 1862, non *G. decurrens* Hook., 1857.

Acrostichum lanceolatum Hook., Sp. Fil. 5 : 276. 1864.

Acrostichum variabile Hook., Sp. Fil. 5 : 277. 1864.

Acrostichum variabile var. *laciniatum* Hook., Sp. Fil. 5 : 277. 1864.

Gymnopteris feei var. *pinnatifida* Beddome, Ferns South India 71. t. 273. 1864.

Leptochilus zeylanicus Fee, Mem. Fam. Foug. 10 : 8. 1865.

Gymnopteris feei var. *trilobata* Beddome, Ferns Brit. India 273. 1866.

Gymnopteris variabilis (Hook.) Beddome, Ferns Brit. India t. 272. 1866 & Handb. Ferns Brit. India 429. 1883.

Gymnopteris lanceolata Beddome, Suppl. Bedd. Ferns South India & Ferns Brit. India 26. 1876.

Gymnopteris variabilis var. *lanceolata* Beddome, Handb. Ferns Brit. India 429. 1883.

Campium lanceolatum (Fee) Copeland, Philipp. Jour. Sci. 37(4) : 348. pl. 5. f. 2. 1928.

Campium decurrens (Bl.) Copeland, Philipp. Jour. Sci. 37(4) : 351. 1928.

Campium zeylanicum (Fee) Copeland, Philipp. Jour. Sci. 37(4) : 352. 1928.

Campium laciniatum Copeland, Philipp. Jour. Sci. 37(4) : 354. pl. 7. 1928.

Leptochilus laciniatus (Hook.) Ching, Bull. Fan Mem. Inst. Biol. 4 : 344. 1933, *quoad syn.* var. *simplex* Ching.

Paraleptochilus decurrens (Bl.) Copeland, Gen. Fil. 198. 1947; Nayar, Amer. Jour. Bot. 50(4) : 301-308. 1963; Nayar & Kaur, Comp. Bedd. Handb. Ferns Brit. India 100. 1974; Satija & Bir, Polyp. Ferns India 81. 1985.

Dendroglossa zeylanica (Fee) Copeland, Gen. Fil. 199. 1947, *quoad syn. Leptochilus decurrens* f. *lanceolatus* (Fee) Sledge, Ann. & Mag. Nat. Hist. Ser. 12(9) : 867. pl. 25. 1956; Manickam & Irudayaraj, Pterid. Fl. W. Ghats 308. pl. 235. 1992, in part; Nayar & Geevarghese, Fern Fl. Malabar 391. 1993.

Leptochilus decurrens f. *laciniatus* Sledge, Ann. & Mag. Nat. Hist. Ser. 12(9) : 870. 1956.
Leptochilus thwaitesianus sensu Nayar & Kaur, Comp. Bedd. Handb. Ferns Brit. India 100. 1974; Chandra & Kaur, Nomen. Guide Bedd. Ferns South India & Ferns Brit. India 6, 25. 1987, in part; Manickam & Irudayaraj, Pterid. Fl. W. Ghats 309. pl. 236. 1992, non Fee, 1865.

(Figs. 21, 22, 23, 25 b, c)

Rhizome moderately to long creeping, 3-4 mm in diameter, palea more dense at the apex and less so in other regions; dictyostelic, vascular bundles 6-10, sclerenchyma strands numerous, bundle sheath paler. Palea ovate-lanceolate or occasionally ovate, 2.5 × 1 mm, dark-brown, clathrate, peltate or basifixed, glandular, margins minutely to distinctly toothed. Fronds dimorphic, in two regular rows on the rhizome, 0.6-1.6 cm apart, articulate; stipe 0.2-18 cm in sterile fronds and 20-26 cm in fertile fronds, occasionally paleate ; sterile fronds : simple, entire, lanceolate, occasionally forked or variously lobed or sub-pinnatifid, 12-30×2.8-6.5 cm, herbaceous or subcoriaceous, apex acuminate, subacute or blunt, base abruptly to gradually decurrent or rounded, margins wavy; fertile fronds : seasonal, filiform, usually with distinct lamina or occasionally lamina-less, simple or sometimes trilobed, 5-11 mm wide, non soriferous marginal area occasionally present. Midrib raised on both surfaces, lateral veins prominent externally or sometimes indistinct, included veins simple or once forked with swollen tip. Sori acrostichoid or sporangia borne directly on the groove of the midrib; paraphysis absent. Annulus 12-14-celled, stalk 3-rowed at the capsular base. Spores 37 × 25 µm, monolete, bilateral, exine spinulose.

Distribution and ecology : *L. decurrens* is distributed in South (Fig. 28) and North-east India to southern China, Sri Lanka, Malesia and Polynesia. It occurs on rocks and base of shrubs and trees near shady streams.

Notes : Blume (1828) described *Leptochilus decurrens* as "L. frondibus simplicibus...longe stipitatis, sterili cuneato oblongo...fertili...elongata" based on a specimen from Java (Fig. 22). Subsequently, Fee (1845) described yet another species *L. lanceolatus* from South India (Fig. 23) as "Frondibus simplicibus...basi decurrentibus...longe petiolatis...leavibus nudis". Many authors like Beddome (1864), Hooker (1864), Copeland (1928, 1947), Ching (1933b), Satija & Bir (1985) and Manickam (1986) treated these two species as distinct. They distinguished *L. decurrens* by its raised lateral veins and lamina gradually decurrent at base as opposed to indistinct lateral veins and lamina abruptly decurrent at base in *L. lanceolatus.* Beddome (1883), contrary to his earlier view (1864) reduced Fee's *L. lanceolatus* as a variety of *Gymnopteris variabilis* Hook. (= *L. decurrens* Bl.) and commented "in typical forms there are no main veins, and all the venation is indistinct, but other forms run too closely into variabilis for it to be considered as a distinct species; the fronds are often quite rounded at base, but at other times nearly as decurrent as in variabilis". Later, Sledge (1956) reduced Fee's *L. lanceolatus* as a forma of *L. decurrens* and remarked "As regards the distinction based on size and shape of frond this is of little significance for whilst typically the fronds of *L. decurrens* are broadly ovate-lanceolate, rather suddenly narrowed below and long decurrent, such plants may commonly be found in Ceylon intermixed in the same population with plants having narrower lanceolate fronds less suddenly narrow

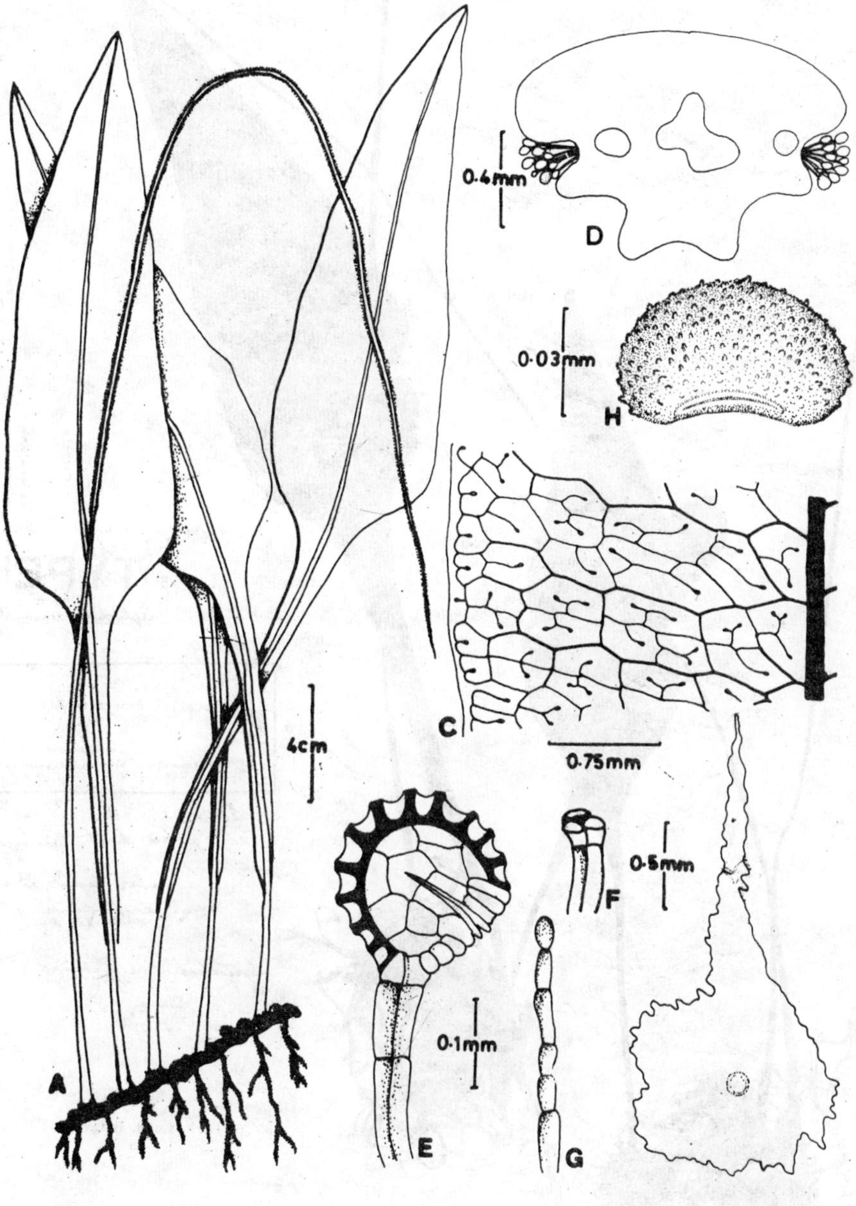

Fig. 21. *Leptochilus decurrens.* **A.** Habit; **B.** Palea; **C.** Venation; **D.** Fertile frond c.s. showing attachment of sporangia; **E.** Sporangium; **F.** Sporangial stalk; **G.** Paraphysis; **H.** Spore. (from *Santhosh Nampy* CU 48611)

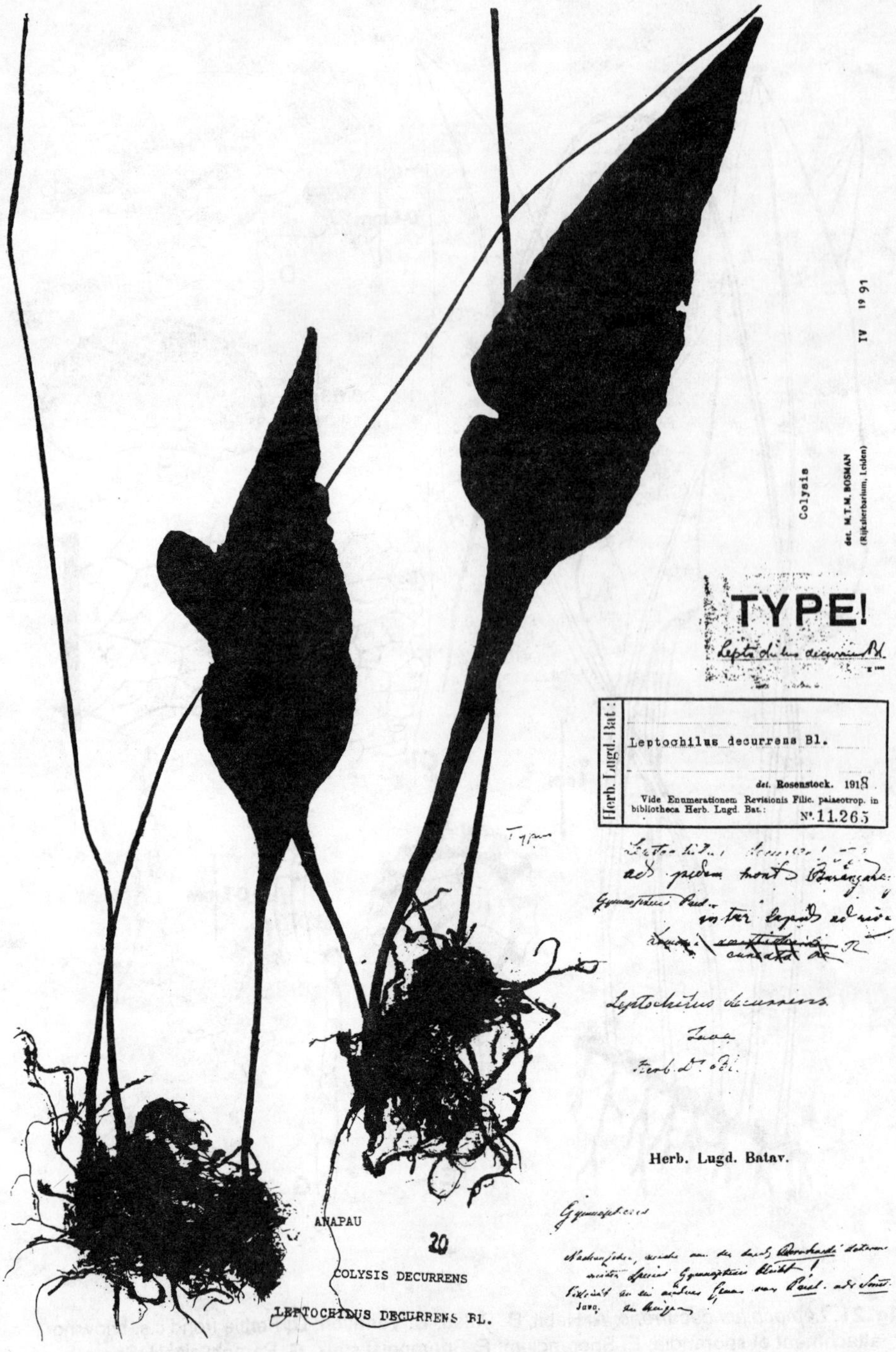

Fig. 22. *Leptochilus decurrens* Blume
(Type : *Blume s.n.*, photocopy)

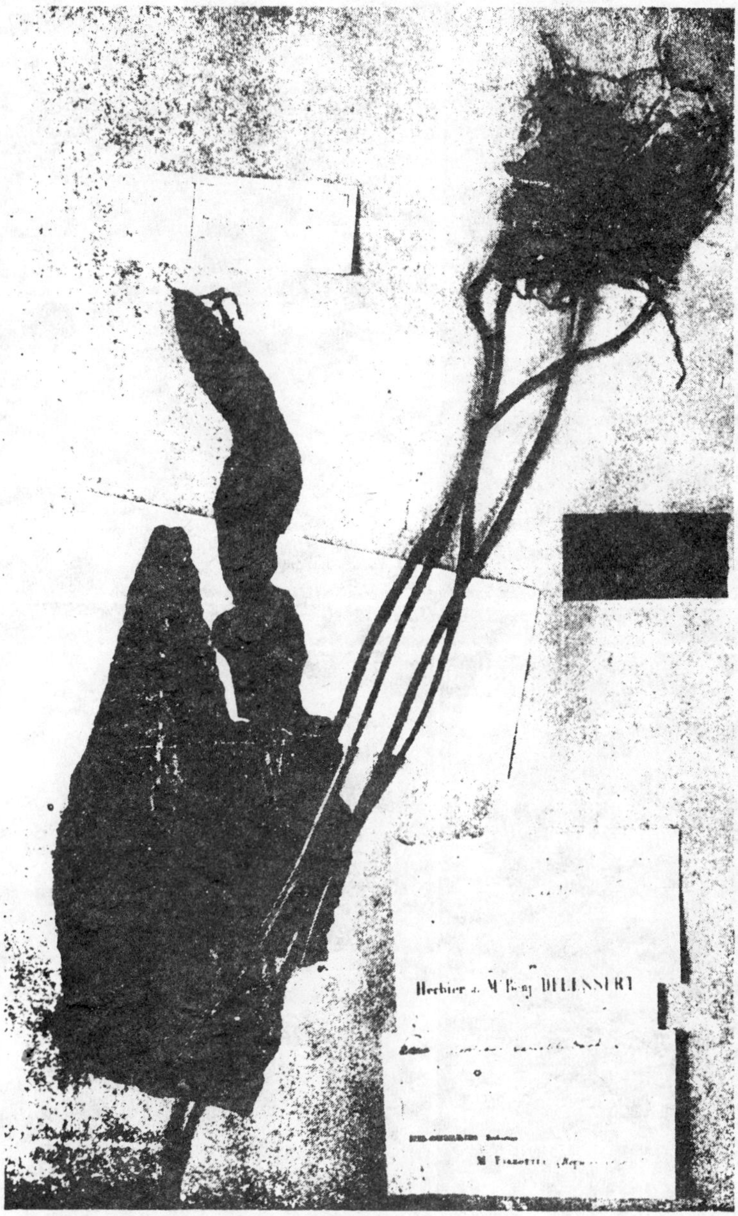

Fig. 23. *Leptochilus lanceolatus* Fee
(Type : *Perottet* 1838, photocopy)

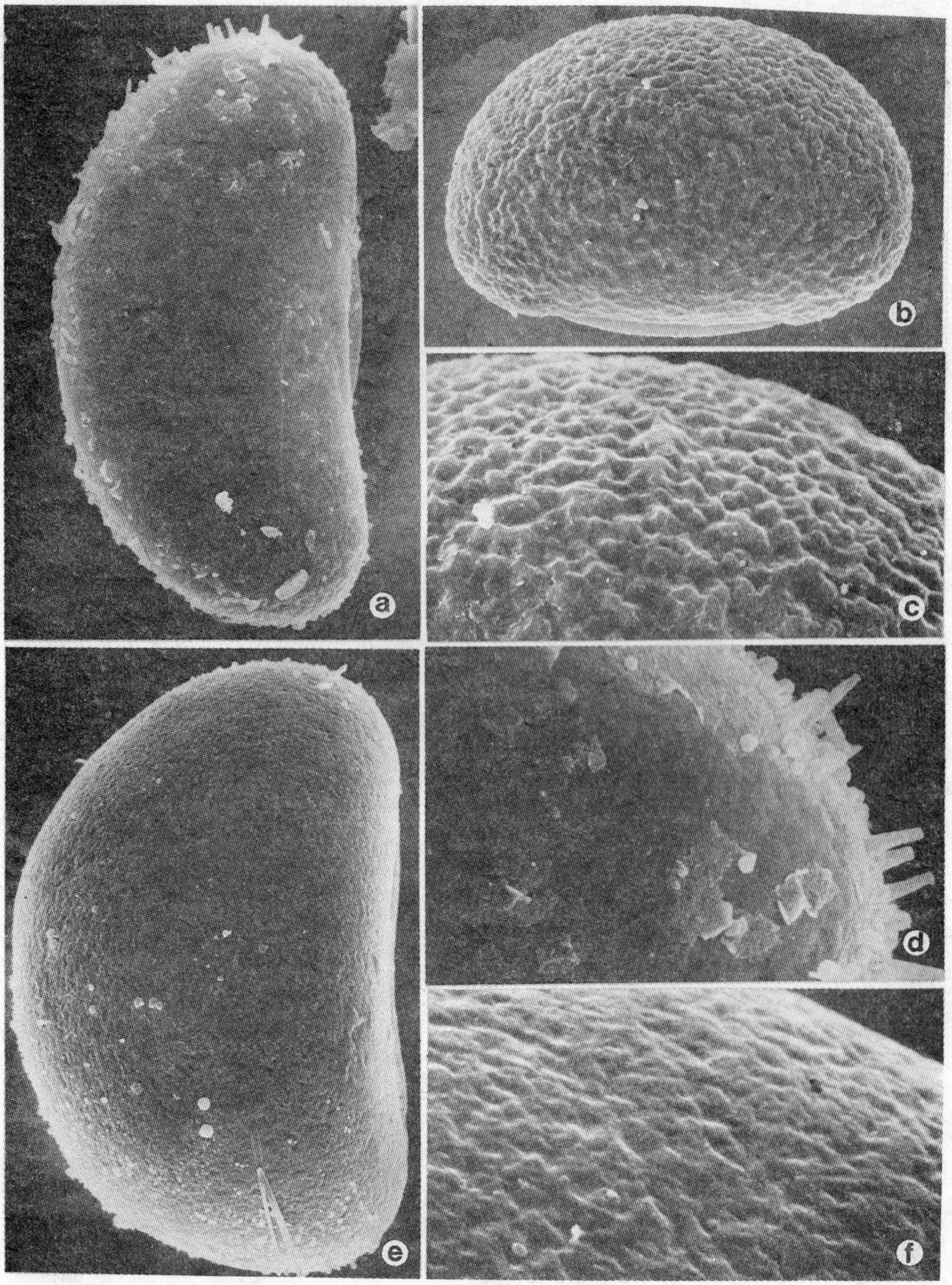

Fig. 24. SEM photograph of spores of polypodioid ferns. **a, d.** *Leptochilus bahupunctika (Santhosh Nampy* 51995); **b, c.** *Lepisorus nudus (Manickam* 1968); **e, f.** *Phymatosorus beddomei (Santhosh Nampy* 52961). **a, b, e.** Lateral view; **c, d, f.** A part enlarged. **a.** x 1500; **b, e.** x 1125; **c.** x 2625; **d.** x 3750; **f.** x 5625.

and long decurrent below. In all other respects they are identical and a continuous series of forms exists between the shapes illustrated by Fee in his two plates". However, he placed the broadly lanceolate forms under forma *decurrens* and narrowly lanceolate forms under forma *lanceolatus*. This was followed by Manickam and Irudayaraj (1992) and Nayar and Geevarghese (1993). After examining a large number of specimens belonging to these species from South India and in consultation with type and authentic specimens, we are inclined to support Sledge (1956) that the characters such as the size and shape of fronds, the venation and texture, etc., are highly plastic and they can not be used for specific recognition at least in this genus so long as they are not correlated with other characters. The original descriptions of these two species deals with only a few characters and they are not much useful to take any taxonomic judgement. Hence, *L. lanceolatus* Fee and *L. decurrens* Blume are treated here as conspecific.

Specimens examined : KARNAṬAKA : Chikmagalur Dt., Bababudan hills, *Santhosh Nampy* 48842 (CALI). Shimoga Dt., way to Agumbe falls, *Santhosh Nampy* 48641; Agumbe, *Santhosh Nampy* 48825; near Agumbe falls, *Santhosh Nampy* 48827 (CALI). KERALA : Idukki Dt., way to Devikulam, *Santhosh Nampy* 48604; Munnar, *Santhosh Nampy* 51999 (CALI). Kozhikode Dt., way to Olichuchattam, *Santhosh Nampy* 52000; Olichuchattam, *Santhosh Nampy* 48855 (CALI). Malappuram Dt., Calicut Univ. Bot. Gard. (Cult.), *Santhosh Nampy* 48652 (CALI). Palakkad Dt., Karimala kovil, *Nayar & Pankaj* 7369; way to Kummattanthode, *Santhosh Nampy* 48653; Kummattanthode, *Santhosh Nampy* 48611; Punnamala, *Nayar* 10622; Silent Valley (submergible area), *Nayar* 10004, *Santhosh Nampy* 48654 (CALI). Pathanamthitta Dt., Anathode, *Nayar* 10112; Kallipara, *Nayar* 10155; Moozhiyar, *Nayar* 10079 (CALI). Thiruvananthapuram Dt., way to Agasthiamala, *Santhosh Nampy* 48610; Athirumala, *Santhosh Nampy* 48655; Ponmudi, *Santhosh Nampy* 45170 (CALI). Wynad Dt., way to Chembra peak, *Santhosh Nampy* 45913; Chembra peak, Madhusoodanan 7263; 8th Hair pin curve, *Geevarghese* 29303; Meppadi, *Geevarghese* 4185 (CALI). TAMIL NADU : Nilgiri Dt., Avalanchi, *Madhusoodanan* 51958, *Santhosh Nampy* 48623; Naduvattam, *Mohanan* 11516; Perar, *Santhosh Nampy* 45165 (CALI). Quaide Milleth Dt., Pillar rocks, *Nayar* 2591; Tiger shola, *Nayar* 2660 (CALI).

Leptochilus metallicus (Bedd.) C. Chr., Ind. Fil. 386. 1906; Sledge, Ann. & Mag. Nat. Hist. Ser. 12 (9) : 876. 1956 & Bull. Brit. Mus. (Nat. Hist.) Bot. 2(5) : 141. 1960; Nayar & Kaur, Comp. Bedd. Handb. Ferns Brit. India 101. 1974; Nayar & Geevarghese, Fern Fl. Malabar 392. 1993.

Gymnopteris metallica Beddome, Suppl. Ferns South India & Brit. India 26. t. 390. 1876 & Handb. Ferns Brit. India 432. t. 260. 1883.

Type : Sri Lanka, *Beddome s.n.* (holotype, BM).

Campium metallicum (Bedd.) Copeland, Philipp. Jour. Sci. 37 (4) : 347. 1928.

Leptochilus zeylanicus sensu Ching, Bull. Fan Mem. Inst. Biol. 4 : 347. 1933, in part, non Fee, 1865.

(Fig. 26)

Rhizome moderately to long creeping, 2-4 mm in diameter, densely paleaceous at the apex and occasionally at older regions; dictyostelic, vascular bundles 4-6,

Fig. 25. Habit of South Indian polypodioid ferns. a. *Leptochilus bahupunctika*; b. *L. decurrens*; c. *L. decurrens* with lobed fronds; d. *Microsorum membranaceum*; e. *M. punctatum*

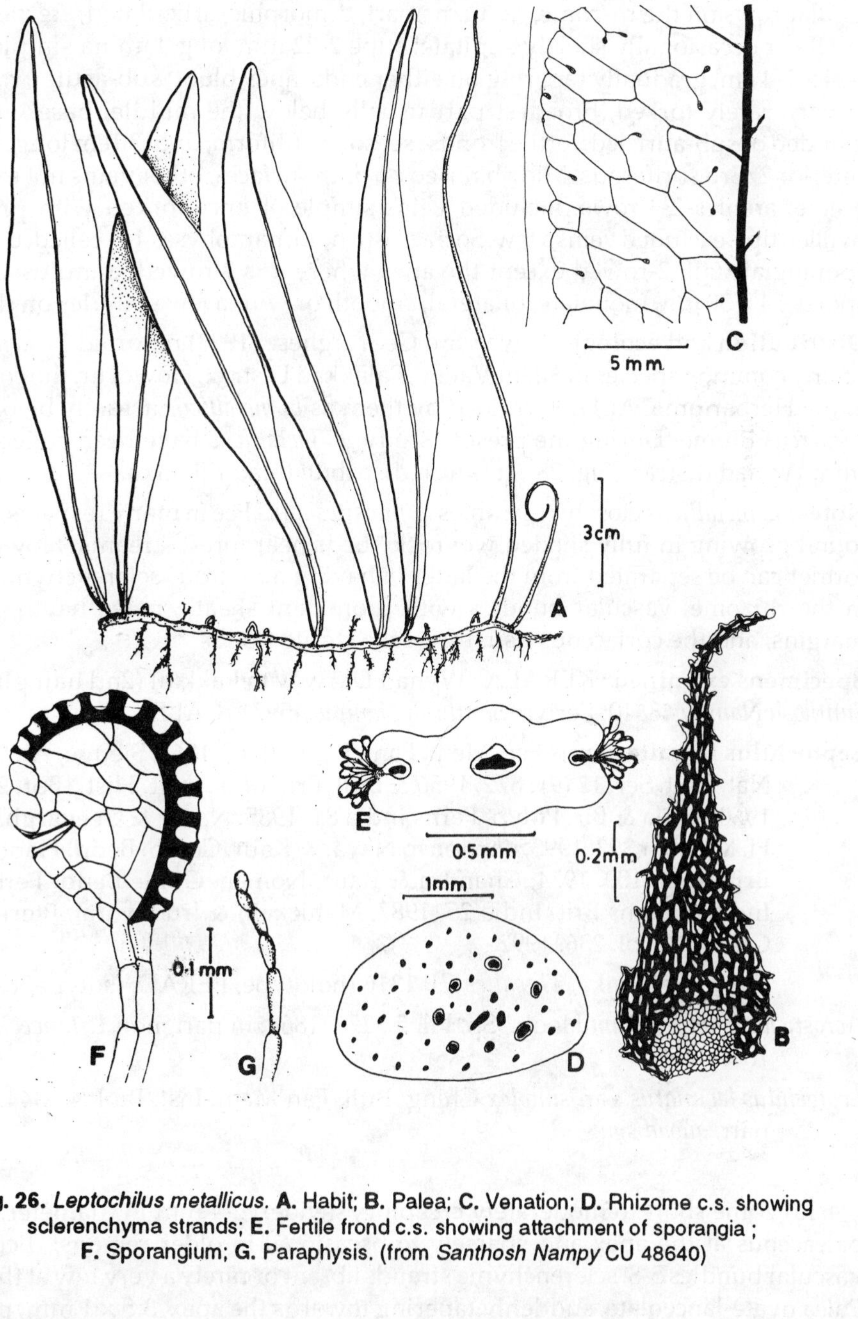

Fig. 26. *Leptochilus metallicus.* **A.** Habit; **B.** Palea; **C.** Venation; **D.** Rhizome c.s. showing sclerenchyma strands; **E.** Fertile frond c.s. showing attachment of sporangia ; **F.** Sporangium; **G.** Paraphysis. (from *Santhosh Nampy* CU 48640)

bundle sheath prominent, sclerenchyma strands numerous. Palea ovate-lanceolate, 2.2×0.5 mm, pseudo-peltate, clathrate, dark-brown, margins toothed. Fronds in two regular rows on the rhizome, *ca.* 1 cm apart, dimorphic, articulated; sterile fronds : sessile or occasionally shortly stipitate; stipe 2-22 mm long; lamina simple, entire, 5-24 × 1-4 cm, gradually tapering on either ends, apex blunt, sub-acute or rounded or very rarely forked, broadest part usually below the middle, base winged or rounded or sub-auricled; fertile fronds : seasonal, filiform, upto 30 cm long, stipitate, anterior 2/3rd soriferous. Midrib raised on both surfaces, other veins not externally visible, areoles 2-3 rows, included veins simple or once forked with prominent swollen tips, excluded veins a few. Sori acrostichoid; paraphyses 4-6-celled, uniseriate. Sporangial stalk 2-rowed except the apex where it is 3-rowed, annulus 14-celled. Spores 54 × 30 µm, monolete, bilateral, smooth or with a few spinules on the exine.

Distribution and ecology : Nayar and Geevarghese (1993) reported *L. metallicus* as a fairly common species in Silent Valley, Palakkad District. However, the collections in the Herbarium CALI determined by them as *L. metallicus* actually belongs to *L. decurrens* Blume. During the present study, *L. metallicus* have been collected only from Wynad district (Fig. 28). It is also distributed in Sri Lanka.

Notes : *L. metallicus* closely resembles *L. thwaitesianus* Fee in many respects, both are found growing in fully shaded wet rocky beds near forest streams. However, the former can be separated from the latter in having numerous sclerenchyma strands in the rhizome, vascular bundles with prominent sheath, palea having toothed margins, and the coriaceous usually sessile sterile fronds.

Specimens examined : KERALA : Wynad Dt., way to Lakkidi (2nd hair pin curve), *Santhosh Nampy* 48640; Periya, *Santhosh Nampy* 45971 (CALI).

Leptochilus thwaitesianus Fee, Mem. Fam. Foug. 10 : 7. 1865; Sledge, Ann. & Mag. Nat. Hist. Ser. 12 (9) : 872. 1956 & Bull. Brit. Mus. (Nat. Hist.) Bot. 2(5) : 141. 1960; Satija & Bir, Polyp. Ferns India 81. 1985; Nayar & Geevarghese, Fern Fl. Malabar 393. 1993, non sensu Nayar & Kaur, Comp. Bedd. Handb. Ferns Brit. India 100. 1974; Chandra & Kaur, Nomen. Guide Bedd. Ferns South India & Ferns Brit. India 25. 1987; Manickam & Irudayaraj, Pterid. Fl. W. Ghats 309. pl. 236. 1992.

Type : Sri Lanka, *Thwaites* CP 1316 (holotype, PERAD; isotype, K, BM).

Acrostichum lanceolatum Hook., Sp. Fil. 5 : 276. 1864, in part, non *L. lanceolatus* Fee, 1865-1868.

Leptochilus laciniatus var. *simplex* Ching, Bull. Fan Mem. Inst. Biol. 4 : 344. 1933, in part, *quoad syn.*

(Fig. 27)

Rhizome short to moderately creeping, slender, 2-4 mm in diameter, densely paleaceous at the apex and sparsely to occasional in older regions; dictyostelic, vascular bundles 5-8, sclerenchyma strands absent or rarely a very few at the centre. Palea ovate-lanceolate, suddenly tapering towards the apex, 3.5 × 1 mm, peltate or pseudo-peltate, clathrate, dark-brown, margins entire, apex and base glandular. Fronds in two dorsal rows on the rhizome, dimorphic, 4-7 mm apart, articulated, stipitate; sterile fronds : stipe 0.5-19 cm, triangular, stramineous to black when dry,

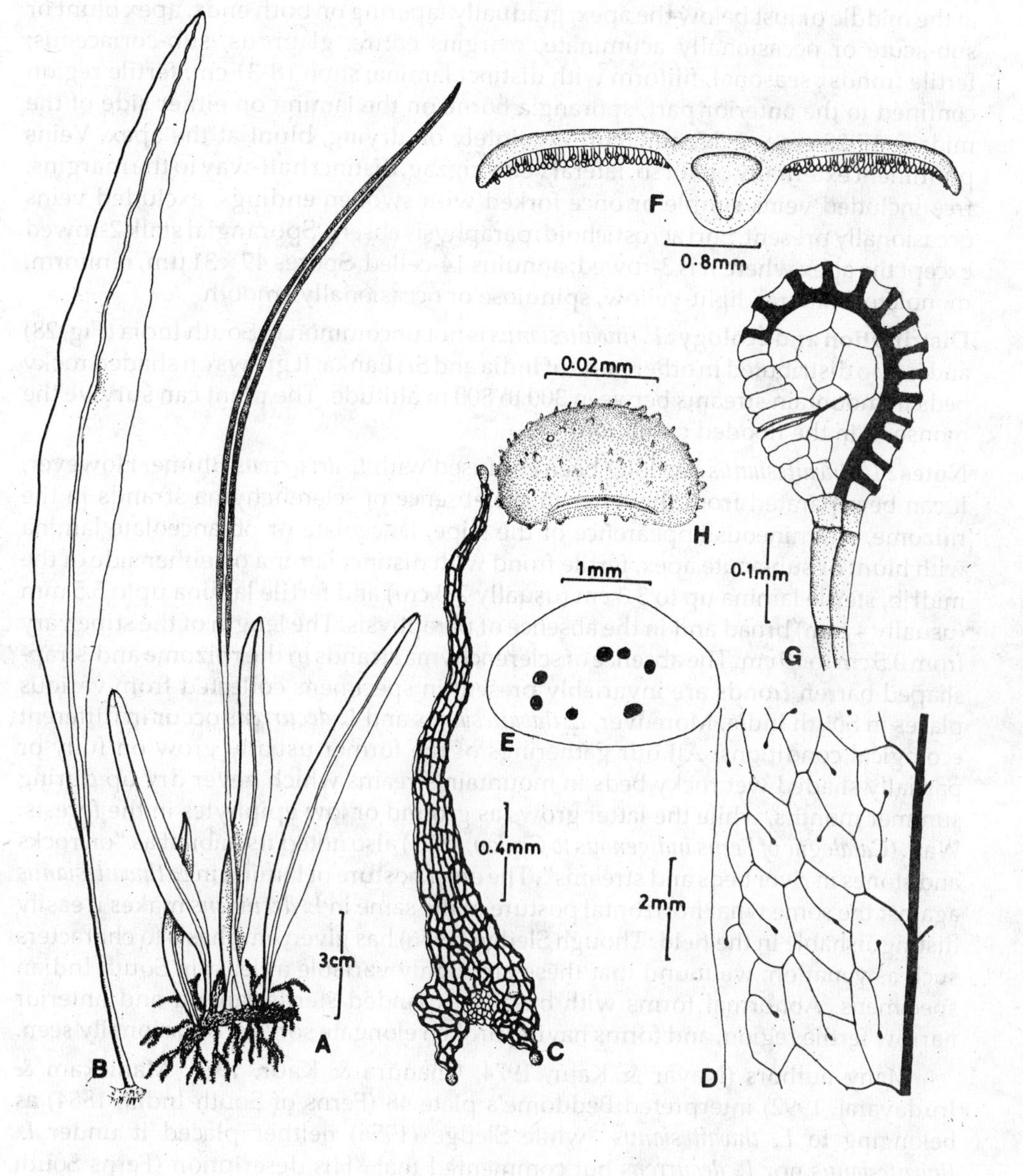

Fig. 27. ***Leptochilus thwaitesianus*. A, B.** **Habit; C. Palea; D. Venation; E. Rhizome c.s. showing absence of sclerenchyma strands; F. Fertile frond c.s. showing attachment of sporangia; G. Sporangium; H. Spore. (from *Santhosh Nampy* CU 45138)**

glabrous; lamina lanceolate or oblanceolate, 9-28 × 1.1-3.3 cm, broadest part usually at the middle or just below the apex, gradually tapering on both ends, apex blunt or sub-acute or occasionally acuminate, margins entire, glabrous, sub-coriaceous; fertile fronds : seasonal, filiform with distinct lamina; stipe 18-31 cm; fertile region confined to the anterior part, sporangia borne on the lamina on either side of the midrib, 15-28 cm × 2-5.5 mm, curl involutely on drying, blunt at the apex. Veins prominent externally or not so, lateral veins zigzag, distinct half-way to the margins, free included veins simple or once forked with swollen endings, excluded veins occasionally present. Sori acrostichoid; paraphysis absent. Sporangial stalk 2-rowed except the apex where it is 3-rowed; annulus 14-celled. Spores 49 × 31 µm, reniform, monolete, bilateral, light-yellow, spinulose or occasionally smooth.

Distribution and ecology : *L. thwaitesianus* is not uncommon in South India (Fig. 28) and is also distributed in other parts of India and Sri Lanka. It grows on shaded rocky beds in mountain streams between 300 to 800 m altitude. The plant can survive the monsoon in the flooded condition.

Notes : *L. thwaitesianus* has long been confused with *L. decurrens* Blume. However, it can be separated from the latter in the absence of sclerenchyma strands in the rhizome, stramineous appearence of the stipe, lanceolate or oblanceolate lamina with blunt or sub-acute apex, fertile frond with distinct lamina on either side of the midrib, sterile lamina up to 3.3 cm (usually 2-3 cm) and fertile lamina upto 5.5 mm (usually 4 mm) broad and in the absence of paraphysis. The length of the stipe vary from 0.5 cm to 19 cm. The absence of sclerenchyma strands in the rhizome and strap-shaped barren fronds are invariably present in specimens collected from various places in South India. Moreover, *L. thwaitesianus* and *L. decurrens* occur in different ecological conditions. All our gatherings of the former usually grow on fully or partially shaded wet rocky beds in mountain streams which never dry up during summer months, while the latter grows as ground or low epiphytes in the forests. Wall. (*Catalogue of Ferns indigenous to Ceylon*, 1873) also noted its habitat as "on rocks and stones in river beds and streams". The erect posture of lamina in *L. thwaitesianus* against the some what horizontal posture of the same in *L. decurrens* makes it easily distinguishable in the field. Though Sledge (1956) has given emphasis to characters such as venation, we found that these are highly variable at least in South Indian specimens. Abnormal forms with basally expanded sterile lamina and anterior narrow fertile region, and forms having broken elongate sori are occasionally seen.

Many authors (Nayar & Kaur, 1974; Chandra & Kaur, 1987; Manickam & Irudayaraj, 1992) interpreted Beddome's plate 48 (Ferns of South India, 1864) as belonging to *L. thwaitesianus* while Sledge (1956) neither placed it under *L. thwaitesianus* nor *L. decurrens* but commented that "His description (Ferns South India 15. t. 48. 1864) suggests *L. thwaitesianus* but the specimen on which the plate was based, and a second sheet also labelled *G. feei* by him, are both atypical plants with abnormally expanded fertile fronds from 5-15 mm broad. The large size and lanceolate acuminate outline of the barren fronds indicate *L. decurrens* rather than *L. thwaitesianus*". We have collected plants looking exactly like Beddome's plate 48 in FSI and the present studies have shown that unlike Sledge (*l.c.*), they all are typical plants and belongs to *L. decurrens*.

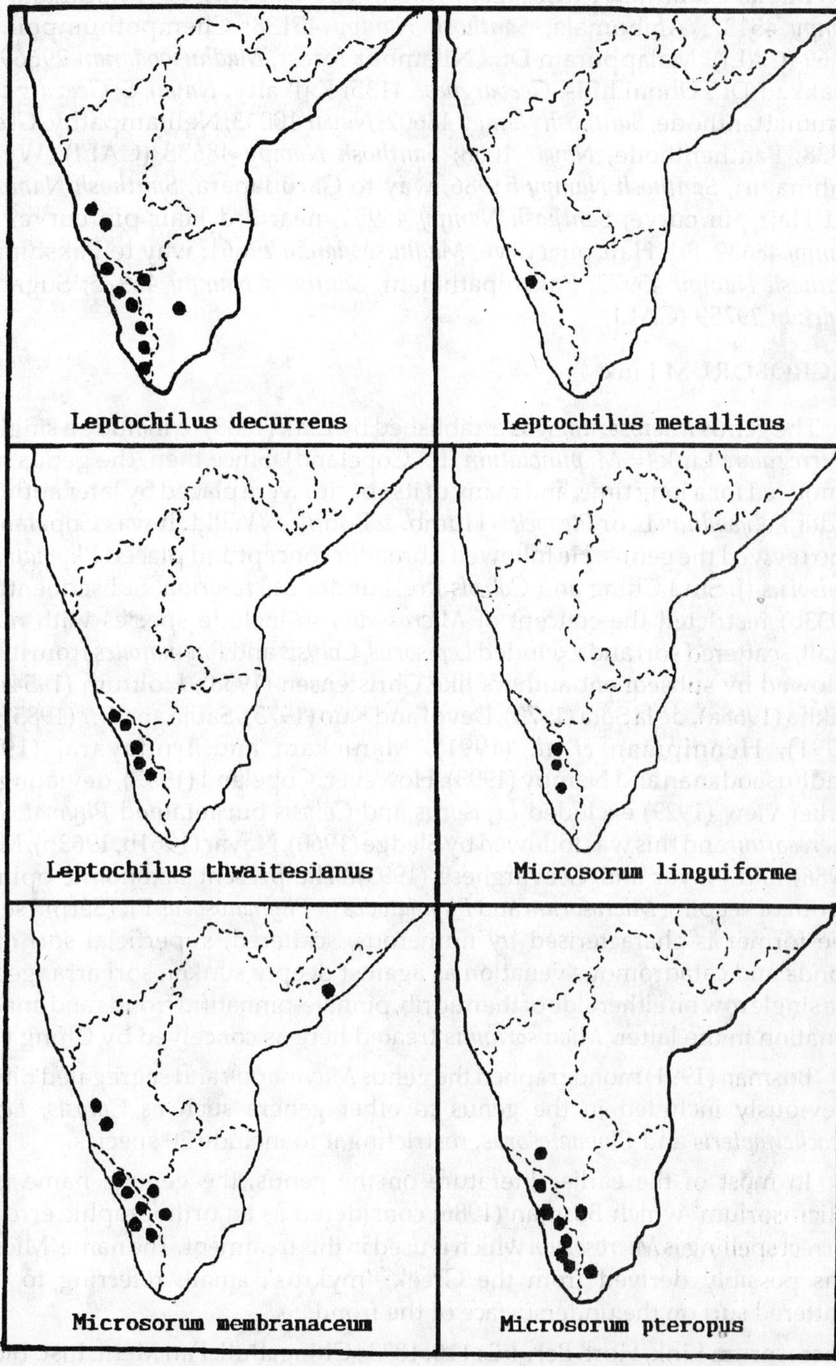

Fig. 28. Distribution maps of species of *Leptochilus* and *Microsorum* in South India

Specimens examined : KERALA : Kozhikode Dt., way to Ambumala, *Santhosh Nampy* 45137; Ambumala, *Santhosh Nampy* 45138; Chempothumpoil, *Pradeep* 51669 (CALI). Malappuram Dt., Nilambur forest, *Madhusoodanan* 29669 (CALI). Palakkad Dt., Dhoni hills, *Geevarghese* 4135; Kaikatty, *Nayar & Geevarghese* 7334; Kummattanthode, *Santhosh Nampy* 48602, *Nayar* 10033; Nelliampathy, *Geevarghese* 17838; Panthenthode, *Nayar* 1068, *Santhosh Nampy* 48638 (CALI). Wynad Dt., Brahmagiri, *Santhosh Nampy* 51986; way to Garudapara, *Santhosh Nampy* 51987; 2nd Hair pin curve, *Santhosh Nampy* 45951; near 3rd Hair pin curve, *Santhosh Nampy* 48639; 8th Hair pin curve, *Madhusoodanan* 29661; way to Pakshipathalam, *Santhosh Nampy* 45972; Pakshipathalam, *Santhosh Nampy* 45975; Sugandhagiri, *Sevichan* 29759 (CALI).

MICROSORUM Link

The genus *Microsorum* was established by Link (1833) to include a single species *M. irregulare* Link (= *M. punctatum* (L.) Copeland). Since then, the genus remained unnoticed for a long time, and many of its species were placed by later authors either under *Polypodium* L. or *Pleopeltis* Humb. & Bonpl. ex Willd. It was Copeland (1929), who revived the genus. He followed a broader concept and placed *Phymatodes* Presl, *Lepisorus* (J. Sm.) Ching and *Colysis* Presl under *Microsorum.* Subsequently, Ching (1933b) restricted the concept of *Microsorum* to include species with numerous, small, scattered sori and excluded *Lepisorus, Colysis* and *Phymatodes* from it. This was followed by subsequent authors like Christensen (1938), Holttum (1954), Bir and Trikha (1968a), de la Sota (1973), Devol and Kuo (1975), Satija and Bir (1985), Bosman (1991), Hennipman *et al.* (1991), Manickam and Irudayaraj (1992) and Madhusoodanan and Nampy (1993). However, Copeland (1947), deviating from his earlier view (1929) excluded *Lepisorus* and *Colysis* but retained *Phymatodes* under *Microsorum* and this was followed by Sledge (1960), Nayar (1961b, 1963b), Manickam (1986) and Nayar and Geevarghese (1993). The present taxonomic opinion is in favour of keeping *Microsorum* and *Phymatodes* (= *Phymatosorus* Pic. Serm.) separately. The former is characterised by numerous, scattered, superficial sori on simple fronds and catadromous venation as against deeply sunken sori arranged usually in a single row on either side of the midrib, pinnate-pinnatifid fronds and anadromous venation in the latter. *Microsorum* is treated here as conceived by Ching (1933b).

Bosman (1991) monographed the genus *Microsorum* and segregated nine species previously included in the genus to other genera such as *Colysis, Leptochilus, Neocheiropteris* and *Phymatosorus,* restricting it to include 20 species.

In most of the earlier literature on the genus, the generic name is spelled 'Microsorium' which Bosman (1986) considered as an orthographic error and the correct spelling is *Microsorum* which is used in this treatment. The name 'Microsorum' was possibly derived from the Greek, 'mykros', small, referring to its small, scattered sori on the undersurface of the frond.

Microsorum Link, Hort. Berol. 2 : 110. 1833 ; Ching, Bull. Fan Mem. Inst. Biol. 4 : 293. 1933; Copeland, Gen. Fil. 195. 1947, in part; Holttum, Fern Fl. Malaya 2 : 170. 1954; Nayar, Bull. Nat. Bot. Gard. Lucknow 58(3) : 1-38. 1961 & Ann. Bot. 27 (105) : 89-100. 1963, both in part; Bir & Trikha, Bull. Bot. Surv. India 10(2) :

133-148. 1968; Devol & Kuo in Li *et al.* (eds.), Fl. Taiwan 1 : 196. 1975; Bosman, Leiden Bot. Ser. 14 : 1-161. 1991 ; Hennipman *et al.* in Kubitzki (ed.), Fam. Gen. Vas. Plants 1 : 219. 1991 ; Manickam & Irudayaraj, Pterid. Fl. W. Ghats 325. 1992; Madhusoodanan & Nampy, Jour. Econ. Tax. Bot. 17(1) : 43-47. 1993; Nayar & Geevarghese, Fern Fl. Malabar 396. 1993, in part.

Type species : *Microsorum irregulare* Link (= *M. punctatum* (L.) Copeland)

Diblemma J. Sm., Jour. Bot. 3 : 399. 1841.

Dendroconche Copeland, Philipp. Jour. Sci. 6 : 96. 1911.

Microsorum sect. *Eumicrosorum* Fee, Gen. Fil. 269. 1852, *nom. inval.*

Rhizome short or long creeping, densely paleaceous in the growing regions. Palea basifixed, pseudo-peltate or peltate, ovate with acuminate apex, clathrate, with an apical and often a few marginal short glandular hairs. Fronds often tufted or distantly placed, simple, entire, usually monomorphous, membranaceous to coriaceous, usually herbaceous, both surfaces often glabrous; stipe articulated. Venation distinct or more or less immersed, main lateral veins distantly placed, and often distinctly raised abaxially and somewhat less adaxially, secondary veins catadromous, smaller veins variously anastomosing, included veins simple, once or twice forked with swollen tips. Sori numerous, irregularly scattered on smaller anastomosing and free included veins, superficial, small, round; with or without paraphysis. Annulus 13-16-celled. Spores monolete, bilateral, smooth to slightly sculptured.

Gametophyte : Two major types of gametophytes are found among the microsorioid ferns : the heart-shaped prothalli (*M. linguiforme, M. membranaceum* and *M. punctatum*) and the ribbon shaped perennial prothalli as in *M. pteropus* (Bosman, 1991).

Chromosome numbers : n = 36, 37 (cf. Hennipman *et al.*, 1991); n = 72 (Manickam & Irudayaraj, 1988).

Distribution and ecology : The distribution of *Microsorum* ranges from tropical and subtropical Africa, through India and South-east Asia to the Pacific. No species of microsorioids are distributed in the Neotropics (Bosman, 1991). Bir and Trikha (1968a) recorded eight species for India. In the present treatment four species of *Microsorum* are described. Most of its species grow in wet and shaded areas as low epiphytes, lithophytes or terrestrials, often in or along streams. *M. punctatum* and a few other species have thick, leathery leaves and grow in more exposed and drier situations.

Key to the species

1a The stipe and midrib hairy below ;
plant usually aqautic lithophyte...............*M. pteropus*

1b The stipe and midrib not hairy;
plant usually epiphyte 2

2a Lamina membranaceous; spores faintly granulate..*M. membranaceum*

2b Lamina herbaceous or sub-coriaceous; spores with smooth exine 3

3a Rhizome short creeping; veins indistinct; fronds narrow ....*M. punctatum*

3b Rhizome wide creeping; veins distinct; fronds broad ....*M. linguiforme*

Microsorum linguiforme (Mett.) Copeland, Univ. Calif. Publ. Bot. 16 : 116. 1929, as *'linguaeforme'*; Nayar & Madhusoodanan, Bot. Jour. Linn. Soc. 75 : 283-298. 1977; Bosman, Leiden Bot. Ser. 14 : 88. 1991; Madhusoodanan & Nampy, Jour. Econ. Tax. Bot. 17(1) : 44. 1993.

Polypodium linguiforme Mett. *In* F.A.W. Miquel (ed.), Ann. Mus. Bot. Lug. 2 : 22. 1866.

Type : New Guinea, *Zippelius s.n.* (holotype, L).

Pleopeltis xiphias Moore, Gdnrs' Chron. 1 : 331. 1881.

Polypodium annabellae Forbes, Jour. Bot. London 26 : 33. t. 280. 1888.

Polypodium cyclobasis Baker, Kew Bull. 42. 1896.

Polypodium schumannianum Diels *In* Schumann & Lauterbach (eds.), Die Fl. von Central-China 139. t. 3 C, D. 1900.

Pleopeltis linguiformis (Mett.) Alderw., Bull. Dept. Agric. Indes neerl. 27 : 6. 1909.

Pleopeltis dendroconchoides Alderw., Bull. Jard. Bot. Buitenz. Ser. 3 (2) : 165. 1920.

Dendroconche kingii Copeland, Univ. Calif. Publ. Bot. 12 : 407. 1931.

(Fig. 29)

Rhizome long creeping, slender, *ca.* 1.2 cm in diameter, densely paleaceous at the apex and less so in older regions; dictyostelic, sclerenchyma strands absent in the ground tissue. Palea peltate or pseudo-peltate, 5 mm long, broadest below the middle, clathrate except for the basal region, ovate-lanceolate, margins slightly toothed, apex and margins glandular. Fronds distantly placed, simple, oblong; stipe articulated, 1-12 cm long, often the stipe is short and lamina extends as a wing nearly to its base; lamina *ca.* 46 × 11.5 cm, base varying from auriculate, or cordate to narrowly angustate, apex rotundate to acuminate, margins entire, herbaceous. Venation reticulate, midrib and other veins distinct, main lateral veins *ca.* 2 cm apart, and form large areoles which extends 2/3 of the breadth of the frond, beyond this there are 2-3 rows of smaller areoles, secondary veins catadromous, smaller veins variously anastomosing within the areoles, free included veins simple, once or twice forked with swollen endings. Sori round or slightly elongate, numerous, irregularly scattered, either on the smaller anastomosing veins or on free included veins, form small tubercles adaxially; paraphyses uniseriate, 1-3-celled. Annulus 17-celled, sporangial stalk 3-celled at the capsular base. Spores 62 × 37 µm, monolete, bilateral, exine smooth.

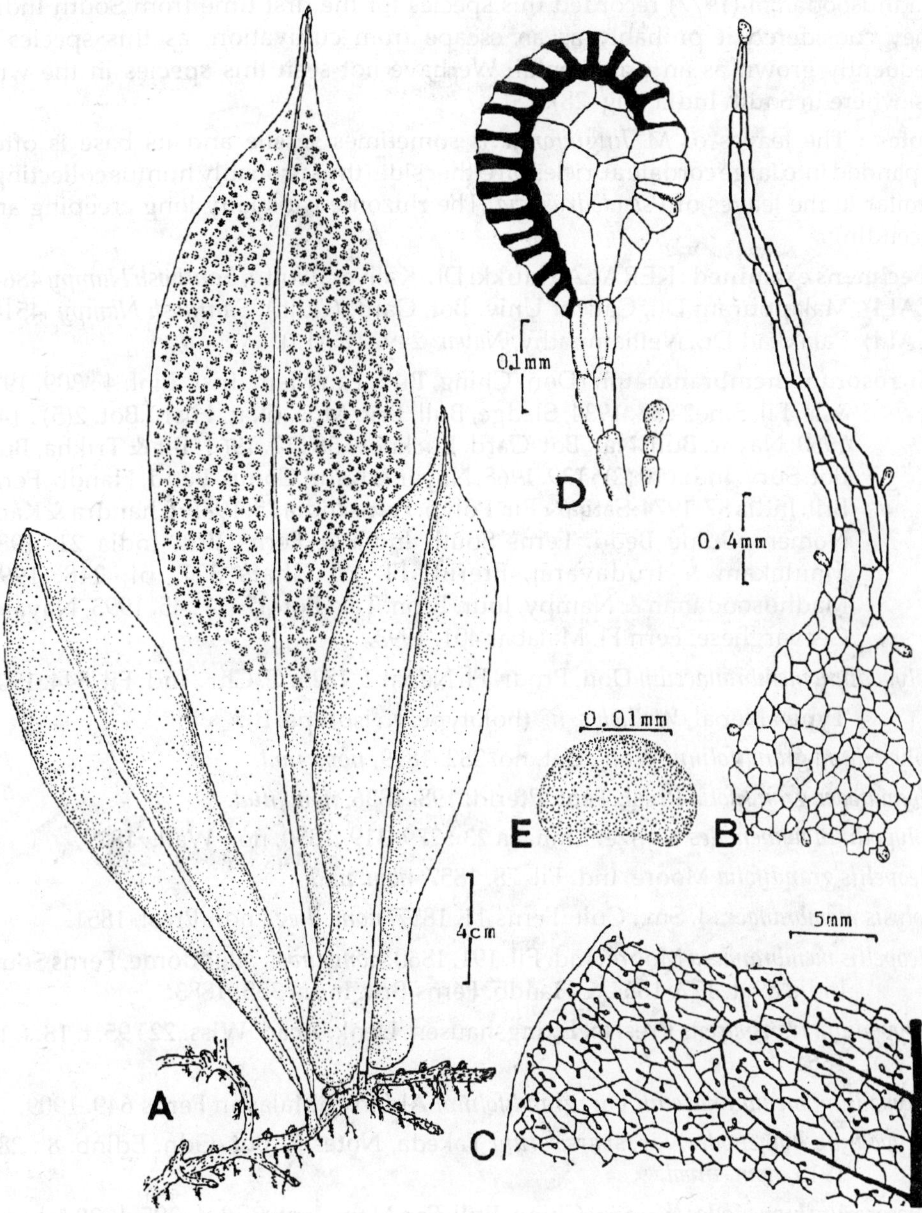

Fig. 29. *Microsorum linguiforme.* **A.** Habit; **B.** Palea; **C.** Venation, **D.** Sporangium; **E.** Spore.
(from *Santhosh Nampy* CU 45182)

Distribution and ecology : *M. linguiforme* is distributed in India, Borneo, Fiji islands, Moluccas, New Guinea, Sumatra, Solomon islands and Sulawesi. Nayar and Madhusoodanan (1977) recorded this species for the first time from South India. They considered it probably as an escape from cultivation, as this species is frequently grown as an ornamental. We have not seen this species in the wild elsewhere in South India (Fig. 28).

Notes : The leaves of *M. linguiforme* is sometimes sessile and its base is often expanded into large cordate auricles on either side (thus probably humus collecting), similar to the leaves of *Pseudodrynaria.* The rhizome is usually long creeping and ascending.

Specimens examined : KERALA : Idukki Dt., Kaliyar (Cult.), *Santhosh Nampy* 48643 (CALI). Malappuram Dt., Calicut Univ. Bot. Gard. (Cult.), *Santhosh Nampy* 45182 (CALI). Palakkad Dt., Nelliampathy, *Nayar* 2492 (CALI).

Microsorum membranaceum (Don) Ching, Bull. Fan Mem. Inst. Biol. 4 : 309. 1933 & Ic. Fil. Sin. 2 : 88. 1934; Sledge, Bull. Brit. Mus. (Nat. Hist.) Bot. 2(5) : 142. 1960; Nayar, Bull. Nat. Bot. Gard. Lucknow 58 : 17. 1961; Bir & Trikha, Bull. Bot. Surv. India 10 (2) : 139. 1968; Nayar & Kaur, Comp. Bedd. Handb. Ferns Brit. India 87. 1974; Satija & Bir, Polyp. Ferns India 75. 1985; Chandra & Kaur, Nomen. Guide Bedd. Ferns South India & Ferns Brit. India 21. 1987; Manickam & Irudayaraj, Pterid. Fl. W. Ghats 327. pl. 248. 1992; Madhusoodanan & Nampy, Jour. Econ. Tax. Bot. 17 (1) : 45. 1993; Nayar & Geevarghese, Fern Fl. Malabar 401. 1993.

Polypodium membranaceum Don, Prodr. Fl. Nepal 2. 1825; C. Chr., Ind. Fil. 544. 1906.

Type : Nepal, *Wallich s. n.* (holotype, K; isotype, B).

Polypodium grandifolium Wall., Cat. no. 282. 1829, *nom. nud.*

Phymatodes grandifolia Presl, Tent. Pterid. 198. 1836, *nom. nud.*

Polypodium hymenodes Kunze, Linnaea 23 : 279, 319. 1850, non Wall., 1829.

Pleopeltis grandifolia Moore, Ind. Fil. 78. 1857, *nom. nud.*

Colysis membranacea J. Sm., Cult. Ferns 11. 1857, *nom. illeg.*, non Presl, 1851.

Pleopeltis membranacea Moore, Ind. Fil. 191. 1860, *nom. inval.* ; Beddome, Ferns South India 59. t. 177. 1864 & Handb. Ferns Brit. India 355. 1883.

Polypodium transparens Presl ex Ettingshausen, Denk. Acad. Wiss. 22 : 95. t. 18. f. 16. 1864.

Polypodium membranaceum var. *grandifolium* Alderw., Malayan Ferns 649. 1909.

Polypodium hymenodes var. *sparsisorum* Takeda, Notes R. Bot. Gdn. Edinb. 8 : 287. 1915, *nom. inval.*

Microsorum hymenodes (Kunze) Ching, Bull. Fan Mem. Inst. Biol. 4 : 295. 1933 & Ic. Fil. Sin. 2 : 84. 1934; Bir & Trikha, Bull. Bot. Surv. India 10 (2) : 139. 1968; Bhargavan, Bull. Bot. Surv. India 15 (3 & 4) : 281. 1973.

(Figs. 25d, 30)

Rhizome short creeping, 3-10 mm in diameter, densely paleaceous at the apex and less so in older regions; dictyostelic, sclerenchyma strands numerous in the

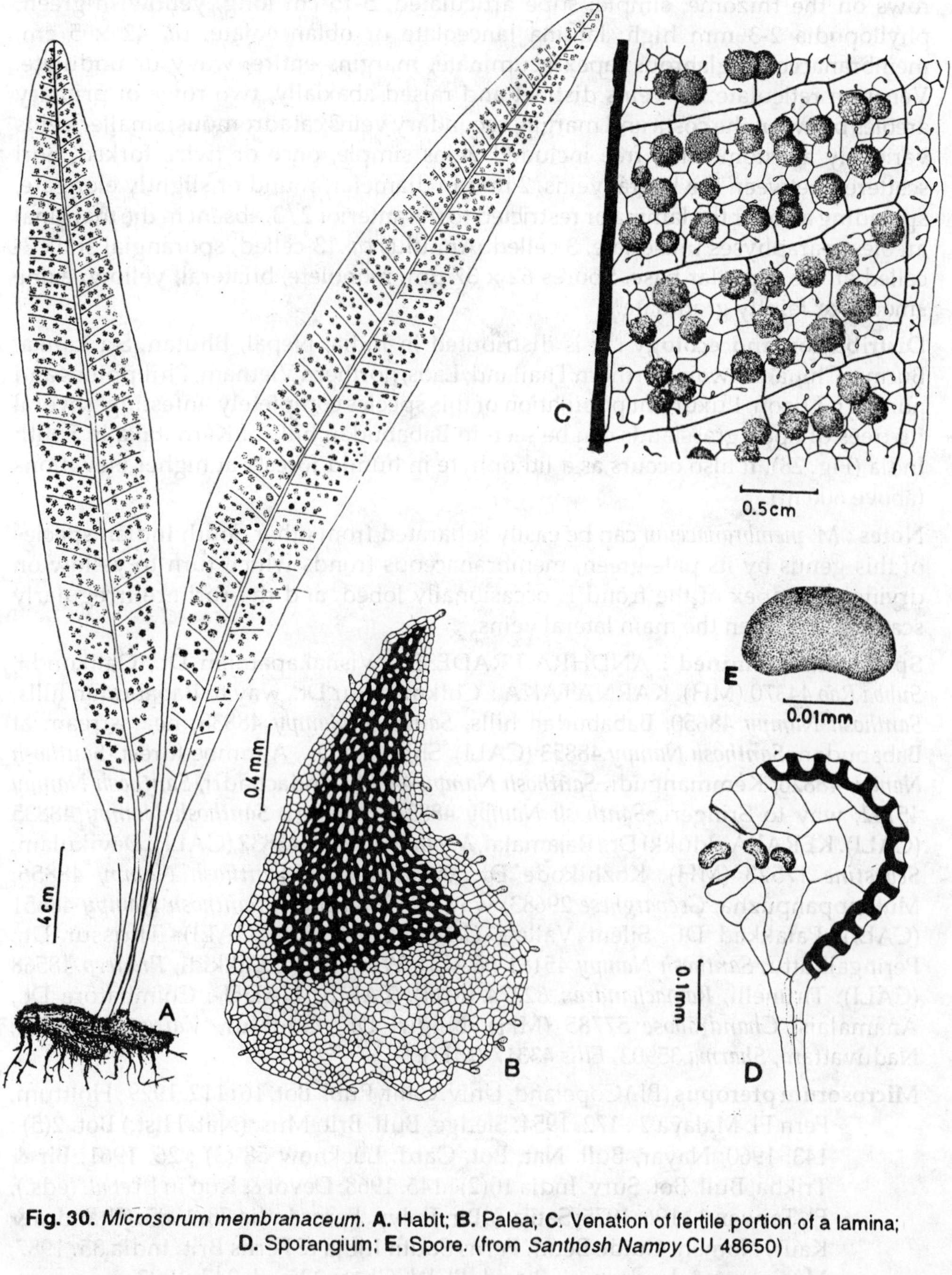

Fig. 30. *Microsorum membranaceum.* **A.** Habit; **B.** Palea; **C.** Venation of fertile portion of a lamina; **D.** Sporangium; **E.** Spore. (from *Santhosh Nampy* CU 48650)

ground tissue. Palea pseudo-peltate or basifixed, ovate, clathrate, central cells thick walled, marginal cells hyaline, apex and margins glandular. Fronds in two dorsal rows on the rhizome, simple; stipe articulated, 5-15 cm long, yellowish-green; phyllopodia 2-3 mm high; lamina lanceolate or oblanceolate, *ca.* 42 × 5 cm, membranaceous, glabrous, apex acuminate, margins entire, wavy or undulate. Venation reticulate, all veins distinct and raised abaxially, two rows of primary areoles between the costa and margin, secondary veins catadromous, smaller veins variously anastomosing, free included veins simple, once or twice forked. Sori scattered between the lateral veins, 2 mm in diameter, round or slightly elongate, spreading all over the lamina or restricted to the anterior 2/3, absent in the marginal areoles; paraphyses uniseriate, 3-celled. Annulus *ca.* 13-celled, sporangial stalk 3-celled at the capsular base. Spores 62 × 37 µm, monolete, bilateral, yellow, exine smooth or faintly granulate.

Distribution and ecology : It is distributed in India, Nepal, Bhutan, Sri Lanka, Burma, China, Taiwan, northern Thailand, Laos, northern Vietnam, Philippines and northern Luzon. Luxurient population of this species completely 'infesting' several fig trees of open grasslands can be seen in Bababudan hills of Karnataka in South India (Fig. 28). It also occurs as a lithophyte in humid forests at higher elevations (above 600 m).

Notes : *M. membranaceum* can be easily separated from other South Indian species of this genus by its pale-green, membranaceous fronds which turn to yellow on drying. The apex of the frond is occasionally lobed, and the sori are irregularly scattered between the main lateral veins.

Specimens examined : ANDHRA PRADESH : Vishakapatnam Dt., Galikonda, *Subba Rao* 44370 (MH). KARNATAKA : Chikmagalur Dt., way to Bababudan hills, *Santhosh Nampy* 48650; Bababudan hills, *Santhosh Nampy* 48838; near stream at Bababudan, *Santhosh Nampy* 48853 (CALI). Shimoga Dt., Agumbe forest, *Santhosh Nampy* 48826; Kemmangudi, *Santhosh Nampy* 44888; Kodachadri, *Santhosh Nampy* 45132; way to Sringeri, *Santhosh Nampy* 48834; Sringeri, *Santhosh Nampy* 48835 (CALI). KERALA : Idukki Dt., Rajamalai, *Madhusoodanan* 43832 (CALI); Devikulam, *Sebastine* 17523 (MH). Kozhikode Dt., Anakampoil, *Santhosh Nampy* 48856; Muthappanpuzha, *Geevarghese* 29683; way to Vellarimala, *Santhosh Nampy* 48651 (CALI). Palakkad Dt., Silent Valley, *Geevarghese* 29320 (CALI). Thrissur Dt., Peringalkuttu, *Santhosh Nampy* 45181 (CALI). Wynad Dt., Lakkidi, *Pradeep* 48568 (CALI); Tirunelli, *Ramachandran* 62774 (MH). TAMIL NADU : Coimbatore Dt., Anamalais, *Chandrabose* 57785 (MH). Nilgiri Dt., Gudalur, *Vajravelu* 41774; Naduvattam, *Sharma* 35903, *Ellis* 43317 (MH).

Microsorum pteropus (Bl.) Copeland, Univ. Calif. Publ. Bot. 16 : 112. 1929; Holttum, Fern Fl. Malaya 2 : 172. 1954; Sledge, Bull. Brit. Mus. (Nat. Hist.) Bot. 2(5) : 143. 1960; Nayar, Bull. Nat. Bot. Gard. Lucknow 58 (3) : 26. 1961; Bir & Trikha, Bull. Bot. Surv. India 10(2) : 145. 1968; Devol & Kuo in Li *et al.* (eds.), Fl. Taiwan 1 : 196. 1975; Satija & Bir, Polyp. Ferns India 78. 1985; Chandra & Kaur, Nomen. Guide Bedd. Ferns South India & Ferns Brit. India 35. 1987; Manickam & Irudayaraj, Pterid. Fl. W. Ghats 326. pl. 247. 1992.

Polypodium pteropus Blume, Enum. Pl. Jav. 125. 1828.

Type : Java, *Blume s.n.* (holotype, L).

Polypodium zosteriforme Wall., Cat. no. 280. 1829, *nom. nud.*

Polypodium tridactylum Wall., Cat. no. 315. 1829, *nom. nud.*

Phymatodes tridactyla (Wall.) Presl, Tent. Pterid. 196. 1836.

Phymatodes zosteriformis Presl, Tent. Pterid. 196. 1836, *nom. nud.*

Drynaria dubia J. Sm. *In* W.J. Hooker (ed.), Jour. Bot. 3 : 397. 1841, *nom. nud.*

Drynaria tridactyla (Wall.) Fee, Mem. Soc. Hist. Nat. Strasbourg 5 : 271. 1852.

Pleopeltis pteropus (Bl.) Moore, Ind. Fil. 78. 1857; Beddome, Handb. Ferns Brit. India 359. 1883.

Pleopeltis tridactyla (Wall.) Moore, Ind. Fil. 78. 1857; Beddome, Ferns South India 60. t. 179. 1864.

Pleopeltis zosteriformis (Wall.) Beddome, Ferns Brit. India t. 123. 1866.

Colysis zosteriformis (Wall.) J. Sm., Hist. Fil. 100. 1875.

Colysis tridactyla (Wall.) J. Sm., Hist. Fil. 101. 1875.

Pleopeltis pteropus var. *minor* Beddome, Handb. Ferns Brit. India 361. t. 204. 1883.

Pleopeltis pteropus var. *zosteriformis* (Wall.) Beddome, Handb. Ferns Brit. India 362. 1883.

Polypodium micropteris Bak., Kew Bull. 14. 1906.

Polypodium aquaticum Christ, Nova Guinea 8 : 153. 1909.

Polypodium udum Christ, Bull. Acad. Int. Geogr. Bot. 20 : 140. 1910.

Microsorum brassi Copeland, Jour. Arnold Arbor. 10 : 181. 1929.

Polypodium pteropus var. *minor* (Bedd.) Wu *et al.*, Bull. Dept. Biol. Sun Yatsen Univ. 3 : 298. pl. 140. 1932.

Microsorum zosteriforme (Wall.) Ching, Bull. Fan Mem. Inst. Biol. 4 : 311. 1933.

Microsorum pteropus f. *minor* (Bedd.) Ching, Bull. Fan Mem. Inst. Biol. 4 : 312. 1933; Chandra & Kaur, Nomen. Guide Bedd. Ferns South India & Ferns Brit. India 21. 1987; Madhusoodanan & Nampy, Jour. Econ. Tax. Bot. 17 (1) : 46. 1993.

Microsorum pteropus var. *minor* Christensen & Tardieu-Blot, Phanerogamia 4 : 194. 1939.

Kaulinia pteropus (Bl.) Nayar, Taxon 13 : 67. 1964; Nayar & Geevarghese, Fern Fl. Malabar 385. 1993.

Kaulinia zosteriformis (Wall.) Nayar & Kaur, Comp. Bedd. Handb. Ferns Brit. India 87. 1974.

Kaulinia pteropus var. *minor* (Bedd.) Nayar & Kaur, Comp. Bedd. Handb. Ferns Brit. India 87. 1974.

Colysis pteropus (Bl.) Bosman, Leiden Bot. Ser. 14 : 112. 1991.

(Fig. 31)

Rhizome long creeping, 2-4 mm in diameter, more or less densely paleaceous all over; dictyostelic, sclerenchyma strands absent in the ground tissue. Palea pseudo-peltate or basifixed 3 mm long, broadly ovate at base, apex acuminate,

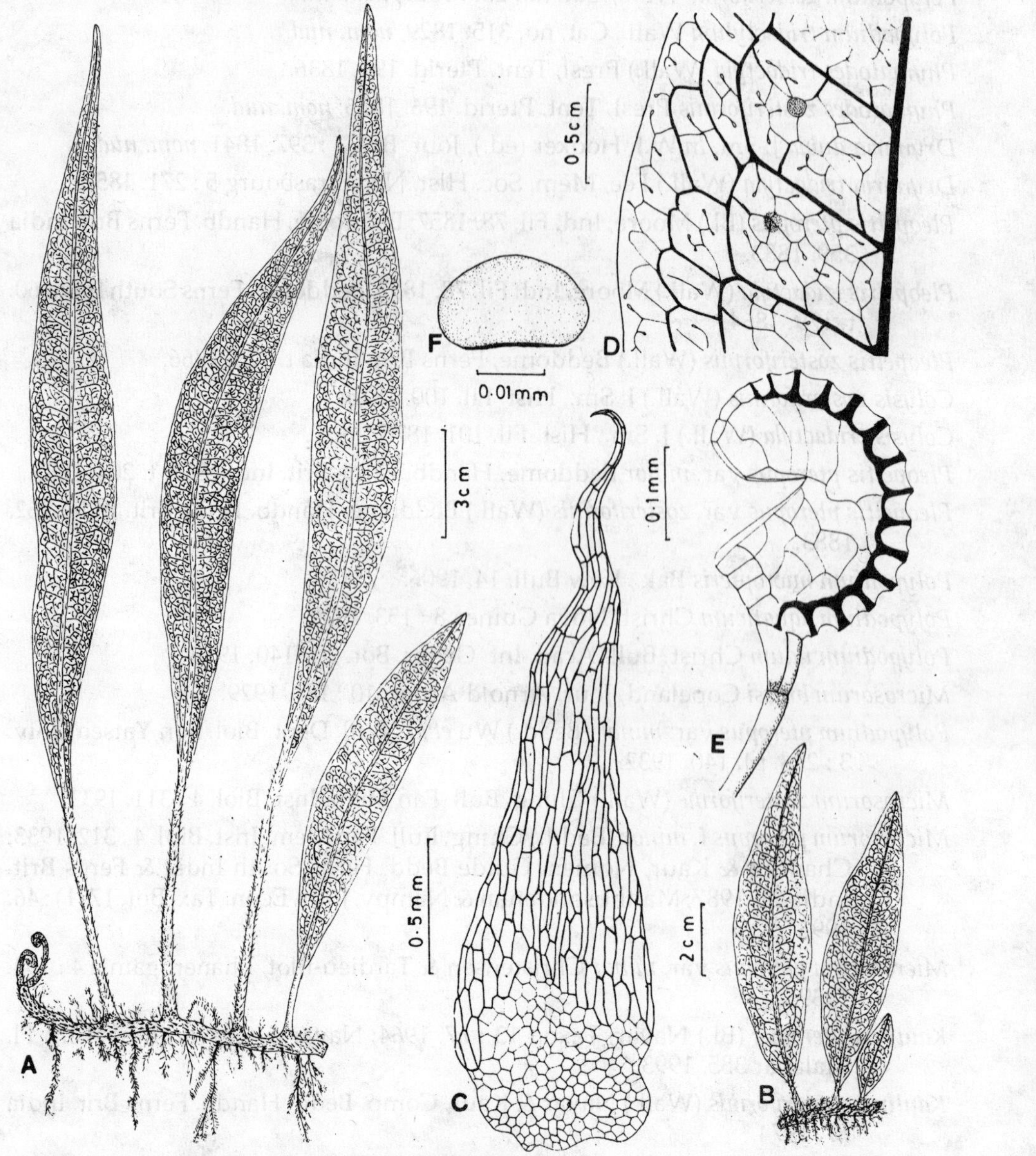

Fig. 31. *Microsorum pteropus.* **A, B.** Habit; **C.** Palea; **D.** Venation of fertile portion of a lamina; **E.** Sporangium; **F.** Spore. (from *Santhosh Nampy* CU 43704)

margins entire, clathrate. Fronds in two dorsal rows on the rhizome, simple, usually 1 cm apart, articulated; stipe 0.5-7 × 1-2 mm, paleaceous all over; lamina lanceolate to oblanceolate, entire or slightly undulate, 6.5-21 × 1.5-2.5 cm, base decurrent, apex acuminate, deep green when fresh, black when dry, herbaceous to membranaceous, surface often raised between the veins, paleaceous over the midrib and main veins. Veins distinct externally, costal areoles large, marginal areoles 1 or 2 irregular rows, secondary veins catadromous, smaller veins variously anastomosing, free included veins simple, once or twice forked with clavate apices. Sori in one row on either side of the midrib, distributed in the main areole, small, round or slightly elongate; paraphyses small, uniseriate, 2 or 3-celled. Annulus 12 or 13-celled. Spores 44 × 26 µm, monolete, bilateral, pale-yellow, exine smooth.

Distribution and ecology : *M. pteropus* is distributed in South India (Fig. 28), North India, Nepal, Burma, Japan, China, Taiwan, Yunnan, Thailand, Laos, Java, Veitnam, Malesia, Sumatra, Borneo, Philippines, Moluccas and New Guinea. It is usually found in wet or damp places above 300 m and grows on rocks, dripping cliffs, or on large roots near forest streams. In rainy season, they remain more or less completely submerged in water.

Notes : The fronds of South Indian specimens pertaining to *M. pteropus* is always simple. However, the fronds of North Indian specimens are usually 3-lobed, while 5-lobed or simple fronds are occasionally met along the same caudex that bears the trilobed fronds (Clarke, 1880; Bir & Trikha, 1968a; Satija & Bir, 1985). *M. pteropus* is characterised by higher density of paleae over the stipe and larger veins, sori distributed in the coastal areole, herbaceous or membranaceous texture of the fronds, absence of sclerenchyma strands in the rhizome and usually lithophytic habit in and along the mountain streams.

Specimens examined : KARNATAKA : Kodagu Dt., Kerati R.F., *Srinivasan* 72315 (MH). KERALA : Idukki Dt., way to Thommenkuttu, *Santhosh Nampy* 48890; Thommenkuttu, *Santhosh Nampy* 43704 (CALI). Kollam Dt., Kulathupuzha, *Sworrupanandan s.n.* (CALI). Kozhikode Dt., Karingad, *Pradeep* 51670; Vellarimala, *Geevarghese* 29471 (CALI). Palakkad Dt., Dhoni, *Geevarghese* 4141; Kanjirapuzha, *Geevarghese* 4055; Nelliampathy, *Geevarghese* 17854; Panthenthode, *Geevarghese* 29254; Silent Valley, *Nayar* 21372, *Madhusoodanan* 1806, *Geevarghese* 17920 (CALI). Pathanamthitta Dt., Moozhiyar, *Nayar* 10104 (CALI). Thiruvananthapuram Dt., Kallar, *Madhusoodanan* 43862 (CALI). Wynad Dt., Chandanathode, *Santhosh Nampy* 48899; Cherunelli, *Nayar & Pankaj* 7345; Periya, *Madhusoodanan* 29744 (CALI). TAMIL NADU : Coimbatore Dt., Anamalais, *Barber* 3674, *Beddome s.n.*; Italiyar R.F., *Ramamurthy* 77523 (MH). Kattabomman Dt., Courtallam, *Sumathykutty Amma* 5984 (CALI).

Microsorum punctatum (L.) Copeland, Univ. Calif. Publ. Bot. 16 : 111. 1929; Ching, Bull. Fan Mem. Inst. Biol. 4 : 309. 1933 & Ic. Fil. Sin. 2 : 86. 1934; Holttum, Fern Fl. Malaya 2 : 179. 1954; Sledge, Bull. Brit. Mus. (Nat. Hist.) Bot. 2(5) : 143. 1960; Nayar, Bull. Nat. Bot. Gard. Lucknow 58 : 19. 1961; Bir & Trikha, Bull. Bot. Surv. India 10 : 143. 1968; Nayar & Kaur, Comp. Bedd. Handb. Ferns Brit. India 87. 1974; Devol & Kuo in Li *et al.* (eds.), Fl. Taiwan 1 : 197. 1975; Satija & Bir, Polyp. Ferns India 77. 1985; Manickam, Fern Fl. Palni Hills 154. 1986; Chandra & Kaur, Nomen. Guide Bedd. Ferns South India & Ferns Brit. India 21. 1987; Bosman, Leiden Bot. Ser. 14 : 97. 1991; Manickam & Irudayaraj,

Pterid. Fl. W. Ghats 328. pl. 249. 1992; Madhusoodanan & Nampy, Jour. Econ. Tax. Bot. 17(1) : 46. 1993; Nayar & Geevarghese, Fern Fl. Malabar 403. 1993.

Acrostichum punctatum L., Sp. Pl. 2 : 1524. 1763.

Type : China, *Fothergill s.n.* (holotype, LINN)

Polypodium polycarpon Cav., Des. de las plantas 246. 1801, non Swartz, 1801.

Polypodium polycarpon Swartz *In* Schrad. (ed.), Jour. Bot. 21. 1801, *nom. illeg.*, non Cav., 1801.

Polypodium irioides Poiret *In* de Lamark (ed.), Botanique 5 : 513. 1804.

Polypodium ambiguum Blume, Enum. Pl. Jav. 125. 1828.

Aspidium microcarpon Blume, Enum. Pl. Jav. 142. 1828, *nom. inval.*

Polypodium polycephalum Wall., Cat. no. 273. 1829, *nom. nud.*

Polypodium glabrum Wall., Cat. no. 281. 1829, *nom. nud.*

Microsorum irregulare Link, Hort. Berol. 2 : 110. 1833.

Polypodium sessile Kaulf. ex Kunze, Bot. Zeit. 6 : 116. 1848.

Pleopeltis irioides (Poiret) Moore, Ind. Fil. 78. 1857; Beddome, Ferns South India 59. t. 178. 1864.

Pleopeltis punctata Beddome, Suppl. Ferns South India & Brit. India 22. 1876 & Handb. Ferns Brit. India 357. 1883.

Polypodium millisorum Bak., Bot. Jour. Linn. Soc. 15 : 109. 1877.

Polypodium superficiale var. *australiense* Bailey, Bot. Bull. Dept. Agric. Queensland 4 : 21.t. 4. 1891.

Polypodium validum Copeland, Ferns 191. 1905.

Polypodium punctatum subsp. *subdrynariaceum* Christ, Bull. Herb. Boissier 994. 1906.

Polypodium punctatum subsp. *subirideum* Christ, Bull. Herb. Boissier 994. 1906.

Polypodium irioides var. *lobatum* f. *cristatum* Bailey, Qd. Agric. Jour. 26(4) : 199. pl. 21. 1911.

Polypodium neoguineense Copeland, Philipp. Jour. Sci. 6 : 89. 1911.

Polypodium aspidistrifrons Hayata, Ic. Pl. 308. f. 103. 1915.

(Figs. 25e, 32)

Rhizome short creeping, *ca.* 1 cm in diameter, densely paleaceous at the apex and less so in older regions; dictyostelic, sclerenchyma strands scattered in the ground tissue. Palea ovate-lanceolate or lanceolate, 5 × 1.5 mm, basifixed or pseudopeltate, clathrate or sub-clathrate, margins dentate to denticulate, apex and margins glandular, glands deciduous. Fronds in two alternating dorsal rows, simple, articulated; stipe short, upto 0.5 cm long; lamina oblanceolate-elongated or broadly linear-lanceolate, narrowed gradually to an attenuated base, apex blunt to acuminate or variously lobed, margins entire or occasionally undulate, herbaceous or subcoriaceous, glabrous, deep green or yellowish-green when fresh, turning ugly dark when dry. Venation reticulate, midrib raised on both surfaces, other veins indistinct,

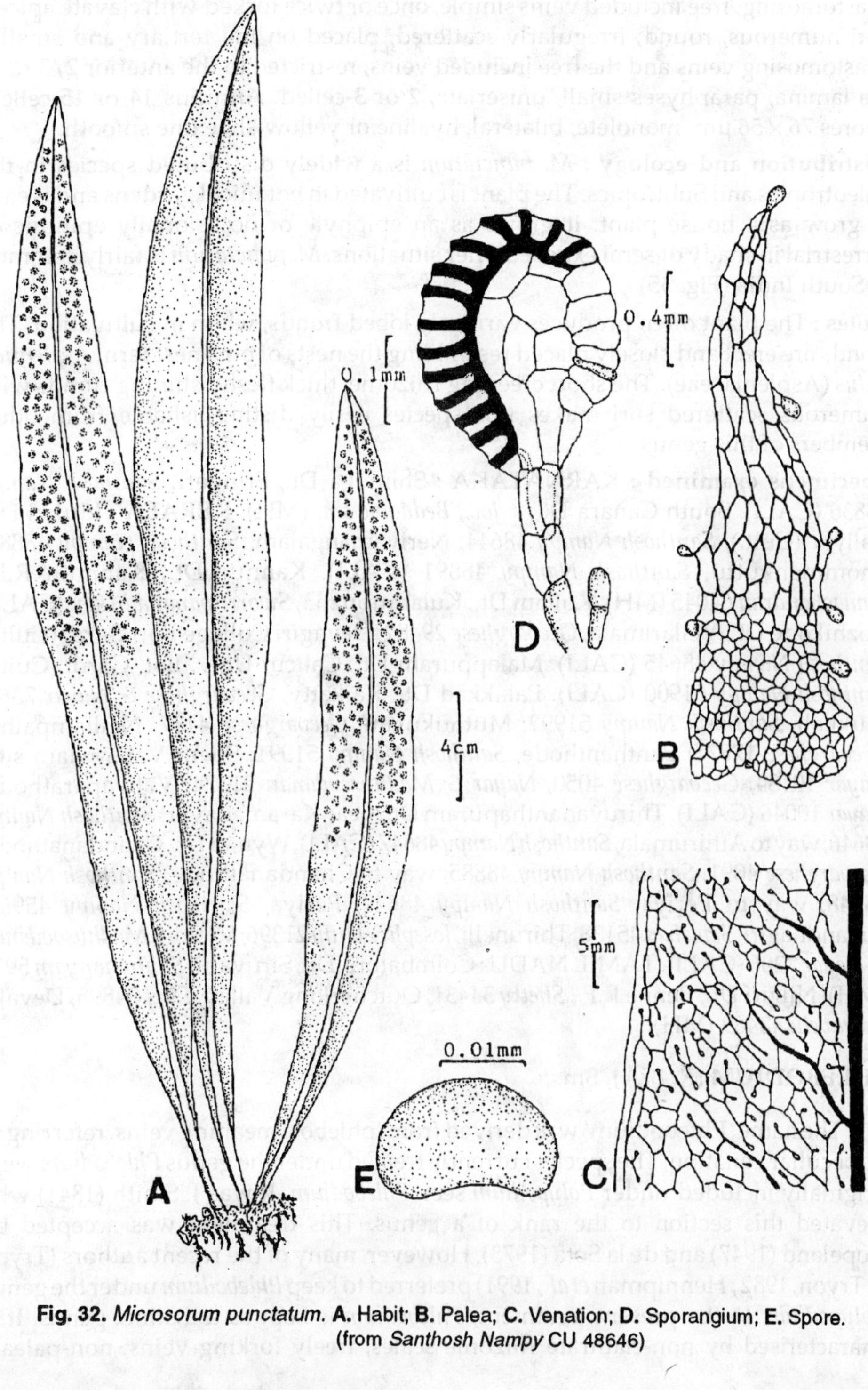

Fig. 32. *Microsorum punctatum*. **A.** Habit; **B.** Palea; **C.** Venation; **D.** Sporangium; **E.** Spore. (from *Santhosh Nampy* CU 48646)

secondary veins catadromous, lateral veins widely separated, smaller veins variously anastomosing, free included veins simple, once or twice forked with clavate apices. Sori numerous, round, irregularly scattered, placed on the tertiary and smaller anastomosing veins and the free included veins, restricted to the anterior 2/3 rd of the lamina; paraphyses small, uniseriate, 2 or 3-celled. Annulus 14 or 15-celled. Spores 76 × 56 µm, monolete, bilateral, hyaline or yellowish, exine smooth.

Distribution and ecology : *M. punctatum* is a widely distributed species in the Paleotropics and Subtropics. The plant is cultivated in botanical gardens and is easy to grow as a house plant. It grows as an epiphyte or occasionally epilithic or terrestrial in shady or semi exposed drier situations. *M. punctatum* is fairly common in South India (Fig. 35).

Notes : The plant often produces variously lobed fronds, when in cultivation. The fronds are erect and closely placed resembling the nests of bird nest fern, *Asplenium nidus* (Aspleniaceae). The short creeping rhizome, thick fleshy shining fronds with numerous scattered sori makes this species easily distinguishable from other members of this genus.

Specimens examined : KARNATAKA : Shimoga Dt., Sringeri, *Santhosh Nampy* 48836 (CALI). South Canara Dt., *s. loc., Beddome s.n.* (MH). KERALA : Idukki Dt., Kaliyar (Cult.), *Santhosh Nampy* 48644; Neriyamangalam, *Santhosh Nampy* 48888; Thommenkuttu, *Santhosh Nampy* 48891 (CALI). Kannur Dt., Kannoth R.F., *Ramachandran* 64045 (MH). Kollam Dt., Kulathupuzha, *Sworrupanandan s.n.* (CALI). Kozhikode Dt., Vellarimala, *Geevarghese* 29482; Devagiri College Bot. Gard. (Cult.), *Santhosh Nampy* 48645 (CALI). Malappuram Dt., Calicut Univ. Bot. Gard. (Cult.), *Santhosh Nampy* 51900 (CALI). Palakkad Dt., Kaikatty, *Geevarghese & Nayar* 7360; Mukkali, *Santhosh Nampy* 51992; Muthukulam, *Geevarghese* 4157; Nelliampathy, *Geevarghese* 17871; Panthenthode, *Santhosh Nampy* 51991; Silent Valley dam site, *Nayar* 10184, *Geevarghese* 4050, *Nayar & Madhusoodanan* 21374; Valiyaparathode, *Nayar* 10046 (CALI). Thiruvananthapuram Dt., near Karamanayar, *Santhosh Nampy* 48646; way to Athirumala, *Santhosh Nampy* 48647 (CALI). Wynad Dt., Chandanathode, *Geevarghese* 4091, *Santhosh Nampy* 48885; way to Chandanathode, *Santhosh Nampy* 48648; way to Periya, *Santhosh Nampy* 48649; Periya, *Santhosh Nampy* 45969; Sugandhagiri, *Sevichan* 45138; Thirunelli, *Joseph & Unni* 21396; Vythiri, *Madhusoodanan & Nayar* 7206 (CALI). TAMIL NADU : Coimbatore Dt., Siruvani, *Subramanyam* 5913 (MH). Nilgiri Dt., Benne R.F., *Shetty* 34431; Outcherlong Valley, *Ellis* 34895; Devala, *Vajravelu* 43469 (MH).

PHLEBODIUM (R. Br.) J. Sm.

The name 'Phlebodium' was derived from 'phlebos' meaning veins, referring to its peculiar venation. The species currently treated under the genus *Phlebodium* were originally included under *Polypodium* sect. *Phlebodium*. It was J. Smith (1841) who elevated this section to the rank of a genus. This treatment was accepted by Copeland (1947) and de la Sota (1973). However, many of the recent authors (Tryon & Tryon, 1982; Hennipman *et al.*, 1991) preferred to keep *Phlebodium* under the genus *Polypodium*. In the present treatment, *Phlebodium* is kept as a distinct genus. It is characterised by non-clathrate rhizome scales, freely forking veins, non-paleate fronds, sori in one row on either side of the midrib, and prominently verrucate or coarsely tuberculate spores.

Phlebodium (R. Br.) J. Sm., Jour. Bot. 4 : 58. 1841; Copeland, Gen. Fil. 187. 1947; Pic. Serm., Webbia 9 (2) : 431. 1953; Dixit & Vohra, Dict. Pterid. India 35. 1984.

Type species : *Phlebodium aureum* (L.) J. Sm.

Polypodium sect. Phlebodium R. Br., Horsf. Pl. Jav. Rar. 4. 1838.

Rhizome long creeping, stout, dictyostelic, densely paleaceous. Palea ovate-lanceolate, apex long acuminate, margins irregularly ciliate. Fronds pinnatifid with wings on either side of the rachis; stipe articulated, shiny; segments lanceolate, glabrous, papyraceous, margins wavy. Veins anastomosing, a row of narrow costal areoles extends from one vein to another without free included veins. Sori in one row on either side of the midrib, round, without paraphysis. Annulus 12 or 13-celled. Spores monolete, bilateral, tuberculate.

Gametophyte : The adult prothallus of *Phlebodium* is cordate-thalloid, usually broader than long reaching maturity in about 6 to 9 months (cf. Nayar & Kaur, 1971).

Chromosome numbers : n = 37, 74 (Evans, 1963) ; n = 74_{II} + 37_{II} (Abraham *et al.*, 1962).

Distribution and ecology : *Phlebodium* is a small genus of about 10 species widely distributed in tropical America and some species under cultivation in South Asia. In South India, the genus is represented by a single species, *P. aureum*. It grows usually as low epiphyte on trees, or on rocks, soil or on rotted wood.

Phlebodium aureum (L.) J. Sm., Lond. Jour. Bot. 4 : 59. 1841 ; Bhargavan & Nair, Jour. Bombay Nat. Hist. Soc. 77 (3) : 539–540. 1981; Dixit, Cens. Ind. Pterid. 48. 1984; Manickam & Irudayaraj, Pterid. Fl. W. Ghats 319. pl. 244. 1992.

Polypodium aureum L., Sp. Pl. 2 : 1087. 1753.

(Fig. 33)

Rhizome long creeping, 1.5 cm in diameter, densely paleaceous all over; dictyostelic, meristeles numerous. Palea ovate-lanceolate, 10 × 2 mm, peltate, attaching region golden-brown, cells uniformly thin-walled, apex long acuminate, margins irregularly ciliate. Fronds pinnatifid, wings broad (8 mm) on either side of the costa; stipe 35 × 0.3 cm, articulated, pale brown and grooved abaxially, dark-brown and round adaxially, shiny; lamina ovate, segments about 13 pairs, slightly ascending, papyraceous, glabrous. Veins anastomosing, one row of slightly ascending areoles present on either side of the costa, free included veins none, excluded veins present. Sori in one row on either side of the midrib, median, placed one in each major areolae, round or elliptic, 2 mm across; non-paraphysate. Annulus 12 or 13-celled, stalk 3-rowed at the capsular base. Spores 68 × 43 μm, monolete, bilateral, coarsely tuberculate.

Distribution and ecology : This species is widely distributed in tropical America and is cultivated as an ornamental in some parts of South India (Fig. 35). Bhargavan and Nair (1981) recorded this species for the first time from the Upper Tiger shola of Nilgiri District in Tamil Nadu. However, we have not encountered this species in the wild anywhere from South India. Earlier report of its occurrence in India (Bhargavan and Nair, *l.c.*) is likely to be an escape from cultivation.

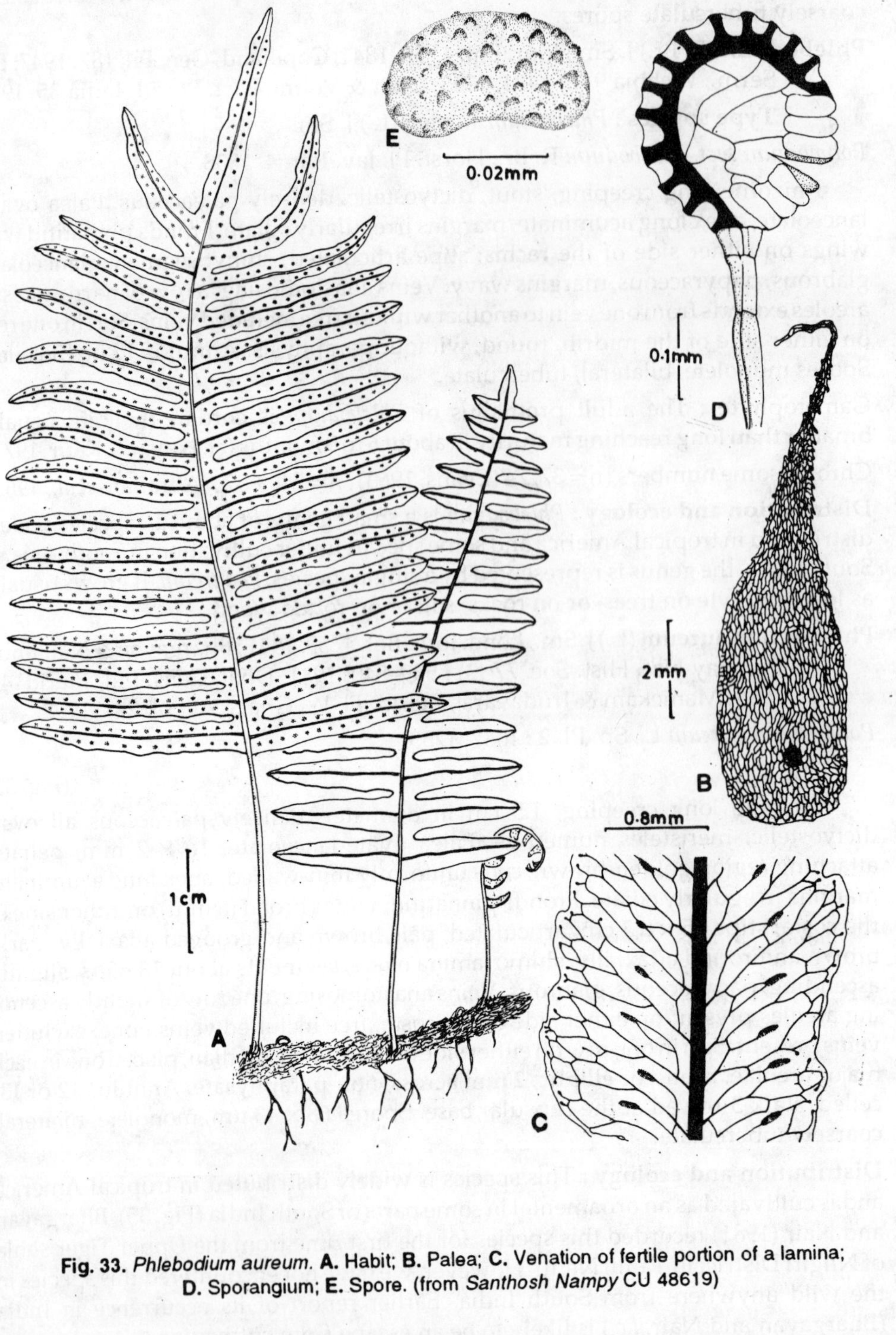

Fig. 33. *Phlebodium aureum*. **A.** Habit; **B.** Palea; **C.** Venation of fertile portion of a lamina; **D.** Sporangium; **E.** Spore. (from *Santhosh Nampy* CU 48619)

Notes : *P. aureum* occasionally produces entire, juvenile leaves and later produces a series of increasingly complex ones. The mature leaves are always pinnatifid.

Specimens examined : KERALA : Idukki Dt., Kaliyar (Cult.), *Santhosh Nampy* 51980 (CALI). TAMIL NADU : Nilgiri Dt., Ooty Bot. Gard. (Cult.), *Santhosh Nampy* 48619 (CALI); Upper Tiger shola, *Sebastine* 4840, *Bhargavan* 47489 (MH).

PHYMATOPTERIS Pic. Serm.

The generic name *Phymatopteris* was first coined by Pichi Sermolli (1973) to include species earlier treated under *Phymatopsis* J. Sm. The genus *Phymatopsis* was originally established by J. Smith in 1875 based on *Polypodium palmatum* Blume and he ascribed 11 species under it. The genus remained unnoticed until it has been revived by H. Ito (1935). However, many contemporary taxonomists like Copeland (1947) and Holttum (1954) considered *Phymatopsis* and another genus *Crypsinus* Presl as congeneric mainly based on the similarity of the non-clathrate rhizome scales and notched margins of the frond but overlooked many other diagnostic characters. Ching (1964) considered these two genera as distinct and states that "the non-clathrate scales and notches in the leaf margin are not a character particular to *Crypsinus* alone, but are characters in common with many different genera within the Polypodiaceae". According to Ching (*l.c.*), the much larger, uniform, simple, trifid, palmatifid, pinnatifid or pinnate leaves of usually chartaceous texture, glaucous or bluish underneath, distinct lateral veins, single rowed but usually superficial or slightly immersed large sori and drynarioid venation are characters peculiar to *Phymatopsis* rather than to *Crypsinus*. On the other hand, *Crypsinus* is small, having dimorphic or sub-dimorphic, rigidly coriaceous, glossy leaves without distinct lateral veins, having a single row of immersed sori on either side of the midrib and with goniophleboid venation. Cytologically these two genera differ, *Phymatopsis* having n = 36 and *Crypsinus* with n = 33. Manton and Sledge (1954) state that "*Crypsinus* of Copeland is polyphyletic and needs generic segregation". Hence *Phymatopsis* and *Crypsinus* are distinct and should be treated separately. *Phymatopsis* J. Sm. is an illegitimate name as it is a later homonym of *Phymatopsis* Tul. ex Trevis., Pichi Sermolli (1973) established the new generic name *Phymatopteris* to replace the former.

The name 'Phymatopteris' refers to the nature of sorus. Here the sorus is slightly immersed, which appears as small tubercles on the upper surface of the lamina. The genus has 13 species in India of which a single species, *P. montana* is represented in South India, while *Crypsinus* is seldom represented in India.

Phymatopteris Pic. Serm., Webbia 28 : 460. 1973.

Type species : *Phymatopteris palmata* (Bl.) Pic. Serm.

Phymatopsis J. Sm., Hist. Fil. 104. 1875, non Tul. ex Trevis., 1857.

Rhizome creeping, densely paleaceous; dictyostelic, sclerenchyma strands conspicuous. Palea peltate, sub-basally attached, lanceolate with a long slender apex, non-clathrate. Fronds simple to pinnate, chartaceous, glaucous or bluish underneath. Venation reticulate, lateral veins distinct. Sori in one row on either side of the midrib, superficial or slightly immersed, large, round; annulus 13 or 14-celled. Spores monolete, bilateral with spinulose exine.

Gametophyte : The prothallus of *Phymatopteris* is strap-shaped for a longer period, but later becomes cordate-thalloid, with a thin midrib and spread out wings (Nayar, 1962).

Chromosome number : n = 36 (Manton & Sledge, 1954).

Distribution and ecology : *Phymatopteris* has about 60 species of which 56 are from the mainland of Asia particularly from China and adjacent countries. Geographically, *Phymatopteris* has its centre of distribution in South-west China and the Himalayas. It grows either on the trunks of tall trees or on moist rocks and cliffs in shaded forests.

Phymatopteris montana (Sledge) Pic. Serm., Webbia 28 : 463. 1973; Satija & Bir, Polyp. Ferns India 63. 1985.

Crypsinus montanus Sledge, Bull. Brit. Mus. (Nat. Hist.) Bot. 2(5) : 145. 1960; Manickam & Irudayaraj, Pterid. Fl. W. Ghats 318. pl. 243. 1992; Nayar & Geevarghese, Fern Fl. Malabar 377. 1993.

Type : Sri Lanka, Adam's peak, *Sledge* 624 (holotype, BM).

Phymatopsis montana (Sledge) Ching, Acta Phytotax. Sin. 9 : 193. 1964.

Pleopeltis oxyloba sensu Beddome, Ferns South India 59. t. 175. 1864, in part, non *Polypodium oxylobum* Wall. ex Kunze, 1851.

Phymatodes montana (Sledge) Bir & Devi, Bull. Bot. Surv. India 10 (2) : 209. 1968.

(Figs. 34, 40a)

Rhizome short creeping, 2-3 mm in diameter, densely paleaceous all over; dictyostelic, sclerenchyma strands numerous. Palea lanceolate, 5 × 1 mm, peltate, sub-basally attached, base broadly ovate, apex gradually tapering into a long, slender, uniseriate hair-like structure, central cells golden-brown, marginal cells paler, margins conspicuously dentate. Fronds in two dorsal rows, 4-10 mm apart, deeply pinnatifid, 4-18 cm long, the number and size of the pinnules vary; pinnules usually 3 or 5 or rarely 7, terminal to be the largest; stipe 1-6 cm long, glabrous, not grooved, brownish, articulated; lamina chartaceous, apex acute, base cuneate, margins nearly smooth or faintly wavy but minutely notched, notches in between adjacent lateral veins. Midrib raised abaxially, lateral veins distinct. Sori in one row on either side of the midrib, 3 mm in diameter, more close to the costa than the margin, placed on a slightly raised receptacle in tertiary or quarternary veins; paraphyses very small, 2 or 3–celled. Annulus 13 or 14–celled; stalk 1.5 times longer than the capsule, 2–rowed except the apex where it is 3-rowed. Spores 80 × 49 µm, monolete, bilateral, golden-brown, exine spinulose.

Distribution and ecology : *P. montana* is distributed in South India (Fig. 35) and Sri Lanka. It usually occurs in exposed rock crevices or as an epiphyte, generally at an altitudinal range 800-2600 m.

Notes : *Phymatopteris montana* was first described by Sledge in 1960 based on a specimen from Sri Lanka. The plant is characterised by inconspicuously ciliate-dentate rhizome scales, smaller fronds (below 30 cm) having fewer and narrower lobes (1-5 pairs and 1-2 cm broad) with minutely notched margins. Sledge (1960) critically examined the South Indian and Sri Lankan plants kept at BM and K and also the specimens used by Beddome for his plate 175 in *Ferns of South India* and commented that "all specimens from South India and Sri Lanka in Herbarium Kew, and Herbarium Brit. Mus. are uniform in their characters and belong to *Crypsinus montanus*" (= *Phymatopteris montana* (Sledge) Pic. Serm.).

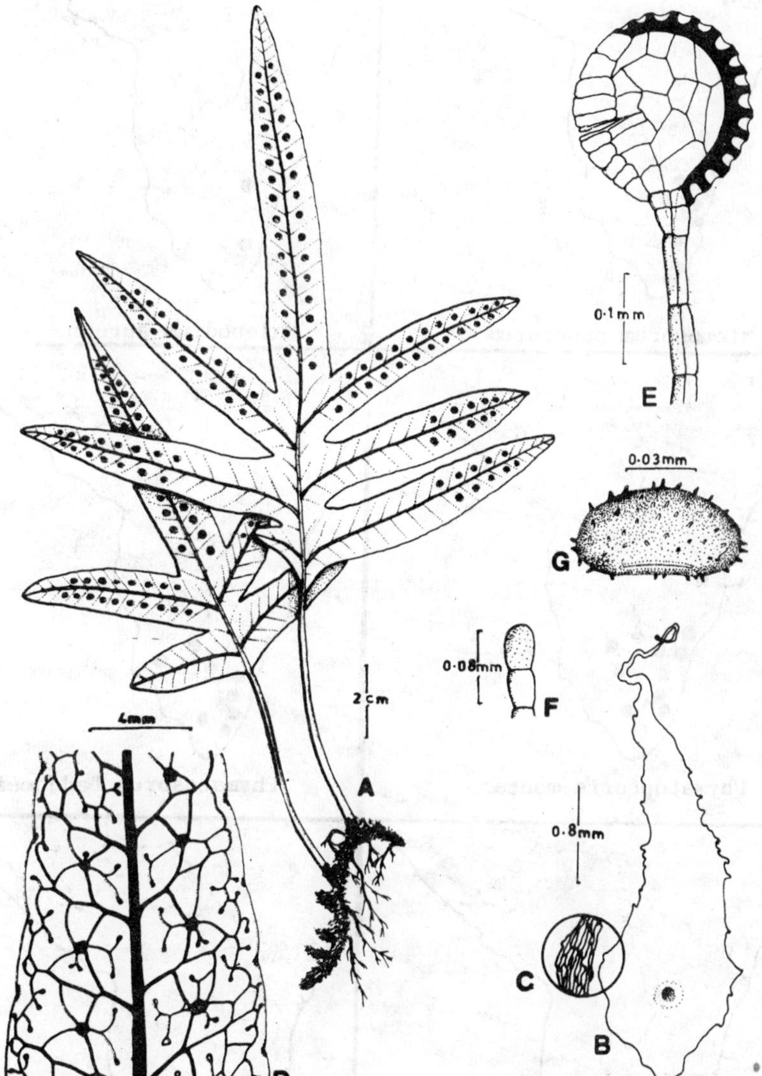

Fig. 34. *Phymatopteris montana.* **A.** Habit; **B.** Palea; **C.** Portion of palea enlarged; **D.** Venation of fertile portion of a lamina; **E.** Sporangium; **F.** Paraphysis; **G.** Spore. (from *Santhosh Nampy* CU 48837)

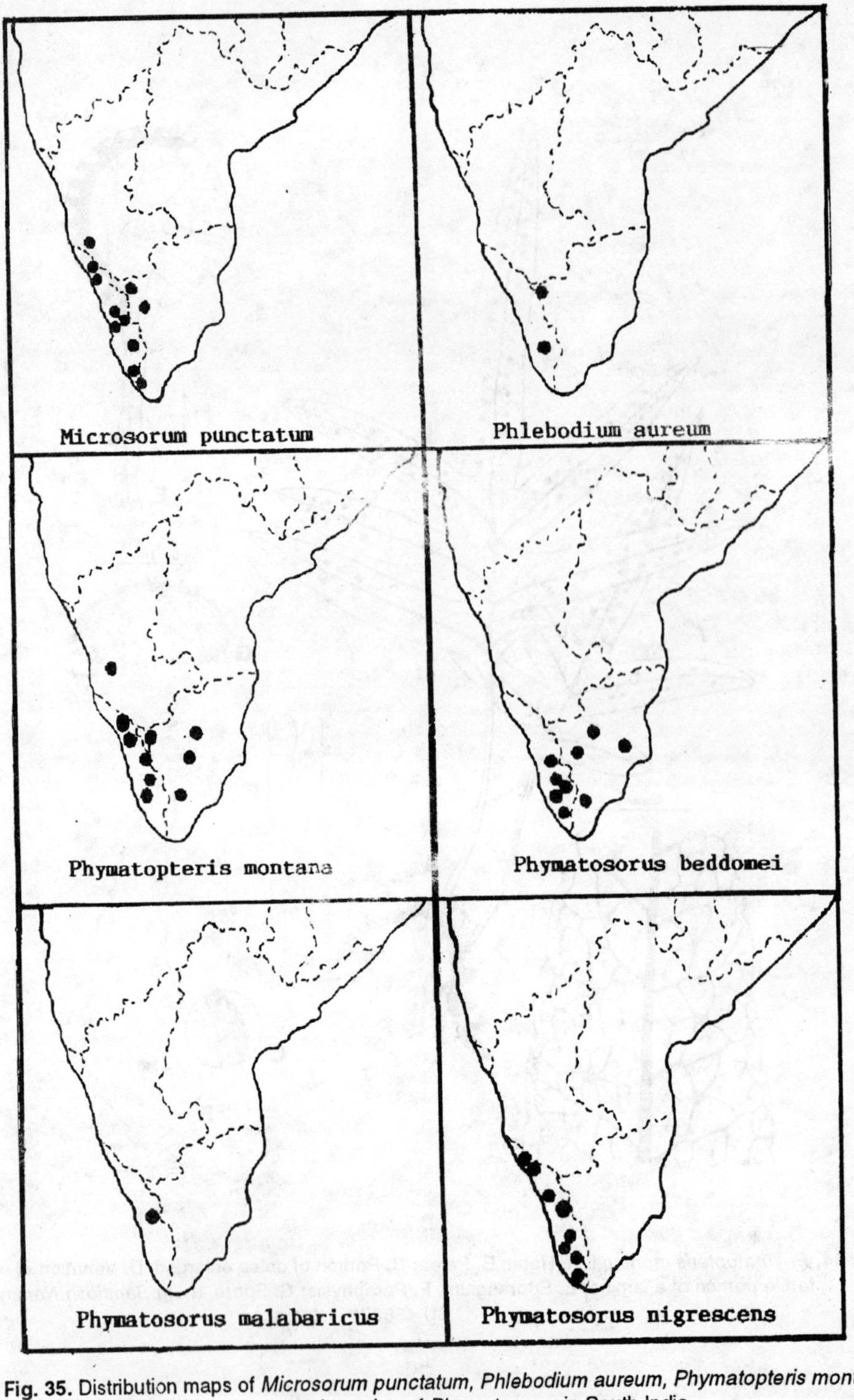

Fig. 35. Distribution maps of *Microsorum punctatum, Phlebodium aureum, Phymatopteris montana* and species of *Phymatosorus* in South India

Specimens examined : KARNATAKA : Chikmagalur Dt., Bababudan hills, *Santhosh Nampy* 48837; way to Bababudan hills, *Santhosh Nampy* 51983 (CALI). KERALA: Idukki Dt., Eravikulam, *Bhargavan* 87324; way to top station, *Deb* 30804; Umaiya Malai, *Shetty* 26634; Vagavarai to Eravikulam, *Bhargavan* 90934 (MH). Kottayam Dt., Lockehrt gap, *Sebastine* 16509; Pamba, *Deb* 30804 (MH). Kozhikode Dt., Olichuchattam, *Santhosh Nampy* 48854 (CALI). Palakkad Dt., Silent Valley, *Nair* 64210 (MH), *Santhosh Nampy & Madhusoodanan* 45985, *Geevarghese* 4039 (CALI). Wynad Dt., Chembra peak, *Geevarghese* 29290 (CALI). TAMIL NADU : Kattabomman Dt., Muthukuzhivayal, *Henry* 47588 (MH). Nilgiri Dt., Bangihalla, *Vivekanandan* 43018; Lakkidi, *Shetti* 34104; Naduvattam, *Subramanyam* 10512, *Sharma* 35919; Parson's Valley, *Ellis* 43448; Pykara, *Vajravelu* 44330, *Narayanaswami* 4300 (MH). Quide Milleth Dt., Bear shola, *Bourne* 4959 (MH), *Santhosh Nampy* 44900 (CALI); Pillar rock stream, *Bourne* 4960 (MH). Salem Dt., Kakasholai hills, *Subramanyam* 7570 (MH).

PHYMATOSORUS Pic. Serm.

Until recently several authors (Smith, 1875; Ching, 1933c, 1964; Holttum, 1954) used the name *Phymatodes* Presl to the genus currently known as *Phymatosorus* Pic. Serm. Copeland (1947), Nayar (1961b), Nayar and Geevarghese (1993), however, considered *Phymatodes* and another genus *Microsorum* as conspecific and used the latter name. Holttum (1954) accepted *Phymatodes* as typified by *P. scolopendria* but synonymised another genus *Lecanopteris* Reinw. under it. Ching (l. c.) divided *Phymatodes* into two sections, namely, sect. *Euphymatodes* and sect. *Phymatopsis.* But, later Ching (1964) changed his view and separated the sect. *Phymatopsis* and restricted the concept of the genus *Phymatodes* to include the sect. *Euphymatodes* (with *Polypodium scolopendria N. L.* Burm. as its type). Morton (1970) has questioned the adoption of the name *Phymatodes* Presl on a nomenclatural ground. He pointed out that according to Article 63, *Phymatodes* is a superflous name and hence illegitimate. Pichi Sermolli (1973) after analysing the problem concluded that "as far as I know, no generic name is available to replace *Phymatodes* as typified by *Polypodium scolopendria* and circumscribed in narrow sense (=*Phymatodes sect. Euphymatodes* Ching)". Consequently, he proposed the new generic name *Phymatosorus,* which is adopted here in the present treatment.

The generic name '*Phymatosorus*' refers to the nature of sori which being sunken and placed in deep cavities that project on the upper surface of the frond as small tubercles.

Phymatosorus Pic. Serm., Webbia 28 : 457. 1973; Satija & Bir, Polyp. Ferns India 68. 1985; Hennipman *et al.* in Kubitzki (ed.), Fam. Gen. Vas. Plants 1 : 221. 1991; Manickam & Irudayaraj, Pterid. Fl. W. Ghats 315. 1992.

Type species : *Phymatosorus scolopendria* (N.L. Burm.) Pic. Serm.

Phymatodes Presl, Tent. Pterid. 195. 1836, *nom. illeg.*; Holttum, Fern Fl. Malaya 2 : 188. 1954, in part; Bir & Devi, Bull. Bot. Surv. India 10(2) : 196-217. 1968, in part.

Phymatodes sect. *Euphymatodes* Ching, Contr. Inst. Bot. Nat. Acad. Peiping 2 : 56. 1993.

Rhizome long creeping, thick, fleshy, sparsely paleaceous in older regions.

Palea peltate, margins and apex glandular, clathrate. Fronds distantly placed, pinnatifid (simple in *P. malabaricus)*; stipe terete, articulated; pinnules lanceolate or linear- lanceolate, glabrous on both surfaces, apex acute or acuminate, coriaceous. Veins anastomosing, tertiary veins anadromous, included veins branched with clavate apices. Sori round, large , in one row (two or more rows in *P. scolopendria*) on either side of the midrib, deeply immersed in cavities, cavities projected towards the upper surface as protuberances; paraphyses simple or with swollen heads. Spores monolete, bilateral, exine smooth.

Gametophyte : Mature prothallus is cordate with marginal and superficial papillate hairs (Nayar, 1961b).

Chromosome numbers : n = 36, 37; 2n = 72, 74 (cf. Hennipman *et al*., 1991).

Distribution and ecology : *Phymatosorus* is a small genus of about 10 species distributed throughout the warmer parts of Asia to Australia, Africa, Madagascar, Mascarenes, Polynesia and in tropical America. Many species of this genus are excellent ornamentals. Majority of the species occur on tree trunks or on moist rocks; some like *P. scolopendria* and *P. nigrescens* grow in exposed conditions. There are four species occur in South India, namely, *P. beddomei* Ghosh, *P. malabaricus* (Nayar & Geev.) Nampy, *P. nigrescens* (Bl.) Pic. Serm. and *P. scolopendria* (N. L. Burm.) Pic. Serm.

Key to the species

1a. Palea ovate-lanceolate with long acuminate apex; sori in two (or more) irregular rows on either side of the midrib; paraphysis absent.................................. *P. scolopendria*

1b. Palea shield like; sori in one row on either side of the midrib ; paraphysis present 2

2a. Rhizome *ca.* 4mm thick, short creeping ; fronds simple .... *P. malabaricus*

2b. Rhizome *ca.* 15 mm thick, long creeping ; fronds pinnatifid 3

3a. Fronds cut into 6-12 segments, membranaceous; veins prominently raised; wing to the rachis broader (5 to 10 mm), pinnules broader (2 to 3 cm); paraphyses with uniseriate stalk bearing two or more irregular rows of balloon shaped cells .... *P. nigrescens*

3b. Fronds cut into more than 12 segments, coriaceous; veins hidden; wing to the rachis below 3 mm, pinnules narrow (*ca.* 1 cm), paraphyses filamentous bearing 6-8 elongate cells .... *P. beddomei*

Phymatosorus beddomei S.R. Ghosh, Jour. Econ. Tax. Bot. 6 : 433-434. 1985.

Type : South India, *H. C. Levinge s. n.* (holotype, CAL).

Pleopeltis leiorhiza sensu Beddome, Ferns South India 58. t. 174. 1864 & Handb. Ferns Brit. India 372. 1883, in part, non Moore, 1862.

Microsorum lucidum sensu Nayar, Bull. Nat. Bot. Gard. Lucknow 58 : 22-24. 1961; Manickam, Fern Fl. Palni Hills 153. 1986, non Copeland, 1947.

Microsorum cuspidatum sensu Nayar & Kaur, Comp. Bedd. Handb. Ferns Brit. India 89. 1974; Nayar & Geevarghese, Fern Fl. Malabar 397. 1993, non Tagawa, 1966.

Phymatodes cuspidata sensu Chandra & Kaur, Nomen. Guide Bedd. Ferns South India & Ferns Brit. India 21. 1987, non J. Sm., 1857.

Phymatosorus lucidus sensu Manickam & Irudayaraj, Pterid. Fl. W. Ghats 316. pl. 241. 1992, non Pic. Serm., 1973. (Fig. 36)

Rhizome long creeping, 1.5 cm in diameter, fleshy, green, densely paleaceous at the apex and less so in older regions; dictyostelic, sclerenchyma strands numerous in the ground tissue. Palea shield-like, *ca.* 2 mm across, peltate, clathrate, margins glandular. Fronds in two dorsal rows, *ca.* 1.5 cm apart, pinnatifid; stipe articulated, *ca* . 56 x 1.5 cm, greenish towards the base and purplish-brown in other regions, terete; pinnules about 7 pairs, sub-opposite or alternate, adnate with the rachis, rachis narrowly winged (2 mm); lamina linear-lanceolate, *ca.* 35 x 1.1 cm, sub-coriaceous, dark-green above and pale-green below, apex acuminate, margins entire or wavy, slightly revolute with conspicuous marginal nerves. Midrib purplish-brown, raised on both surfaces, other veins indistinct, free veins simple or once forked with clavate apices. Sori restricted to the basal 2/3rd part of the lamina, one row on either side of the midrib, nearer to costa than margin, sub-opposite or alternate, seated in cavities which protrudes on the upper surface; paraphysis uniseriate, 6-8-celled. Annulus 14-16-celled; stalk 2-rowed except at the apex where it is 3-rowed. Spores 80 x 59 μm, monolete, bilateral, light yellow, exine smooth.

Distribution and ecology : *P. beddomei* is endemic to South India, and usually occurs at an altitude above 1000 m (Fig. 35). It grows on wet, shaded rocks in forest beds or on the crevices of trees forming dense colonies. The plant is best suited as an ornamental.

Notes : Ghosh (1985) named this species after R.H. Beddome, who collected it from the Anamalais (South India). But Beddome (1864) wrongly attributed the name

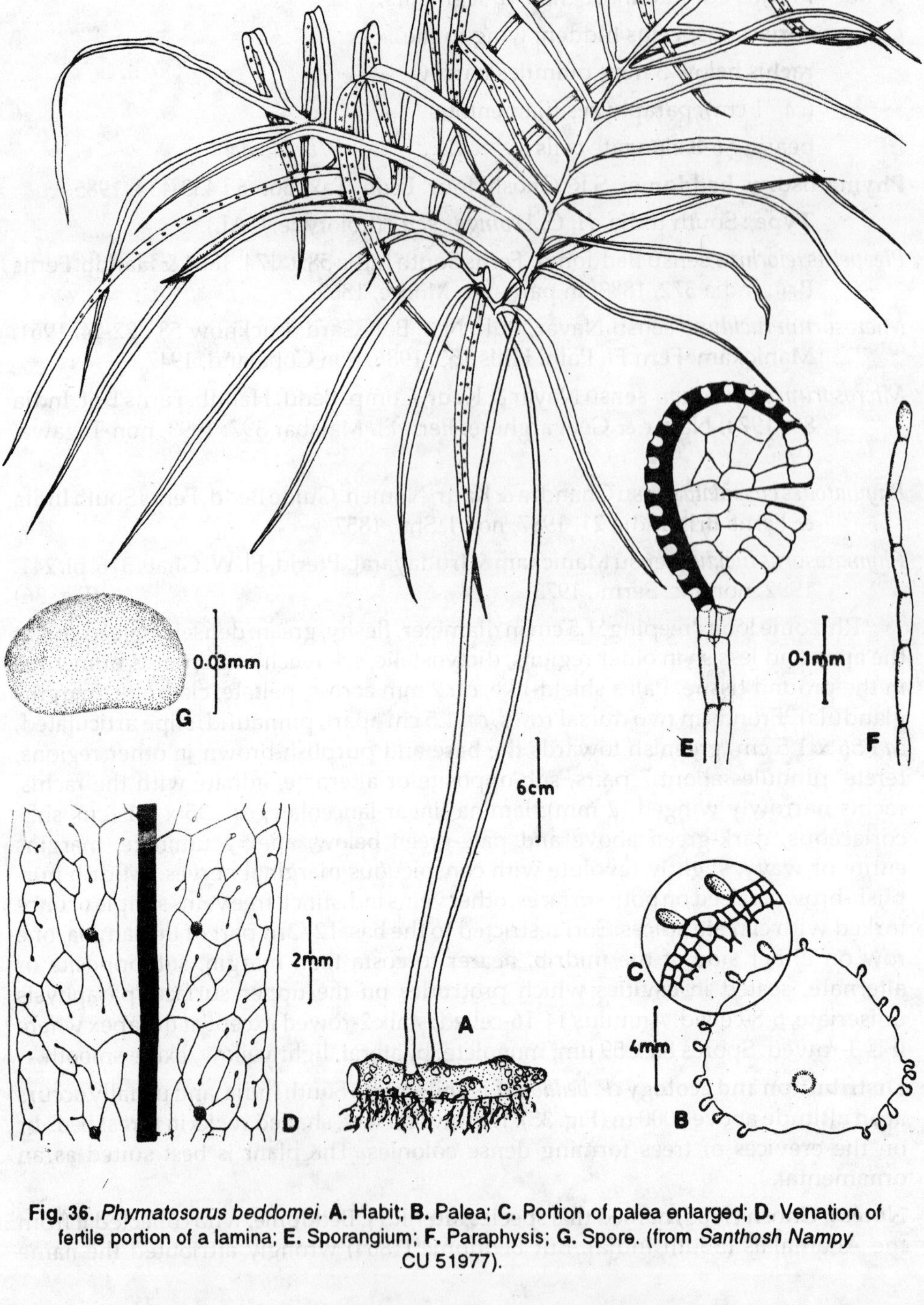

Fig. 36. *Phymatosorus beddomei.* A. Habit; B. Palea; C. Portion of palea enlarged; D. Venation of fertile portion of a lamina; E. Sporangium; F. Paraphysis; G. Spore. (from *Santhosh Nampy* CU 51977).

Pleopeltis leiorhiza Wall. for this species in his *Ferns of South India* (P. 58, pl. 174). *P. leiorhiza* (now *Phymatosorus cuspidatus* (D. Don) Pic. Serm.) is a Himalayan species which differs from the South Indian *Phymatosorus beddomei* (*Pleopeltis leiorhiza* sensu Beddome) in having mostly short petiolate pinnules as opposed to completely sessile pinnules decurrent and adnate with rachis in *P. beddomei.*

Specimens examined: KERALA: Idukki Dt., Kaliyar (Cult.), *Santhosh Nampy* 51977; Munnar, *Madhusoodanan* 43715 (CALI); way to Mangaladevi temple, *Nair* 70228; Pambupara to Vandanmedu, *Vivekananthan* 50560; way to Vazhukkapara, *Nair* 67879 (MH). Kottayam Dt., Pampadumpara, *Subramanyam* 10228; forest near Santhanpara, *Sebastine* 18354 (MH). Palakkad Dt., Karapara river, *Vajravelu* 46136, *Nair* 69653 (MH); Karimalakovil, *Nayar* 7371; Parambikulam, *Geevarghese* 17854 (CALI). Pathanamthitta Dt., Kochupamba, *Nayar* 10131; Moozhiyar, *Nayar* 10161 (CALI). TAMIL NADU: Coimbatore Dt., Attakatti to Valparai, *Balakrishnan* & *Ellis* 11716; Kavarackal, *Chandrabose* 57252 (MH); Valparai, *Santhosh Nampy* & *Madhusoodanan* 51972 (CALI); Waverly estate, *Joseph* 13080 (MH). Kattabomman Dt., Kanthaparai, *Henry* & *Chandrabose* 19848 (MH). Madurai Dt., Aruna estate, *Subramanyam* 9483 (MH). Nilgiri Dt., Ooty Bot. Gard. (Cult.), *Santhosh Nampy* 48620 (CALI). Ramanathapuram Dt., Nagarias estate, *Srinivasan* 68066; Sethur hills, *Srinivasan* 65944 (MH). Salem Dt., Konkapallam, *Nair* 74248 (MH).

Phymatosorus malabaricus (Nayar & Geev.) Nampy *comb. nov.*

Microsorum malabaricm Nayar & Geev arghese, Fern F1. Malabar 399. p1. 163. 1993.

Type: Kerala, *K. K. Geevarghese* 4073 (holotype, CALI).

Phymatosorus malabaricus Geev arghese ex Nampy & Madhusoodanan, Fer Gaz. 14 (7) : 233-234. 1994, *nom. superf.*

Microsorum malabaricm (Geev. ex Nampy & Madhus.) Nooteboom, Blumea 42 (2) : 355. 1997.

(Fig. 37)

Rhizome short creeping, slender, 3-5 mm in diameter, densely paleaceous at the apex. Palea sub-orbicular to ovate, pseudo-peltate with overlapping basal auricles, clathrate, entire, apex and margins glandular. Fronds in two dorsal rows, 1-4 mm apart, simple , entire; stipe 30 x 1 mm, basally paleaceous, articulate; lamina lanceolate, 12-15 x 1.2-1.8 cm, papyraceous, apex acuminate, base cuneate or gradually tapering, broadest at the middle, glabrous. Venation reticulate, midrib raised on both surfaces, free included veins often branched with swollen tips. Sori in one row on either side of the midrib, ovate, 1.5-2 x 1 mm, deeply sunken, almost equidistant to the costa and margin; paraphyses filamentous, stalk uniseriate, 3 or 4-celled, body biseriate with 2 or 3-tiers of short balloon-like cells ending in a swollen cell. Annulus 19-21-celled. Spores 50 x 36 μm, monolete, bilateral, exine smooth.

Distribution and ecology: This species is rather rare and endemic to Kerala, where it occurs on moss covered, wet rocks in forests (Fig. 35).

Notes: *P. malabaricus* is quite distinct from all other South Indian species of this genus in having short, slender rhizome and simple, entire fronds.

Specimen examined : KERALA : Palakkad Dt., Kanjirapuzha, *Geevarghese* 4073 (CALI).

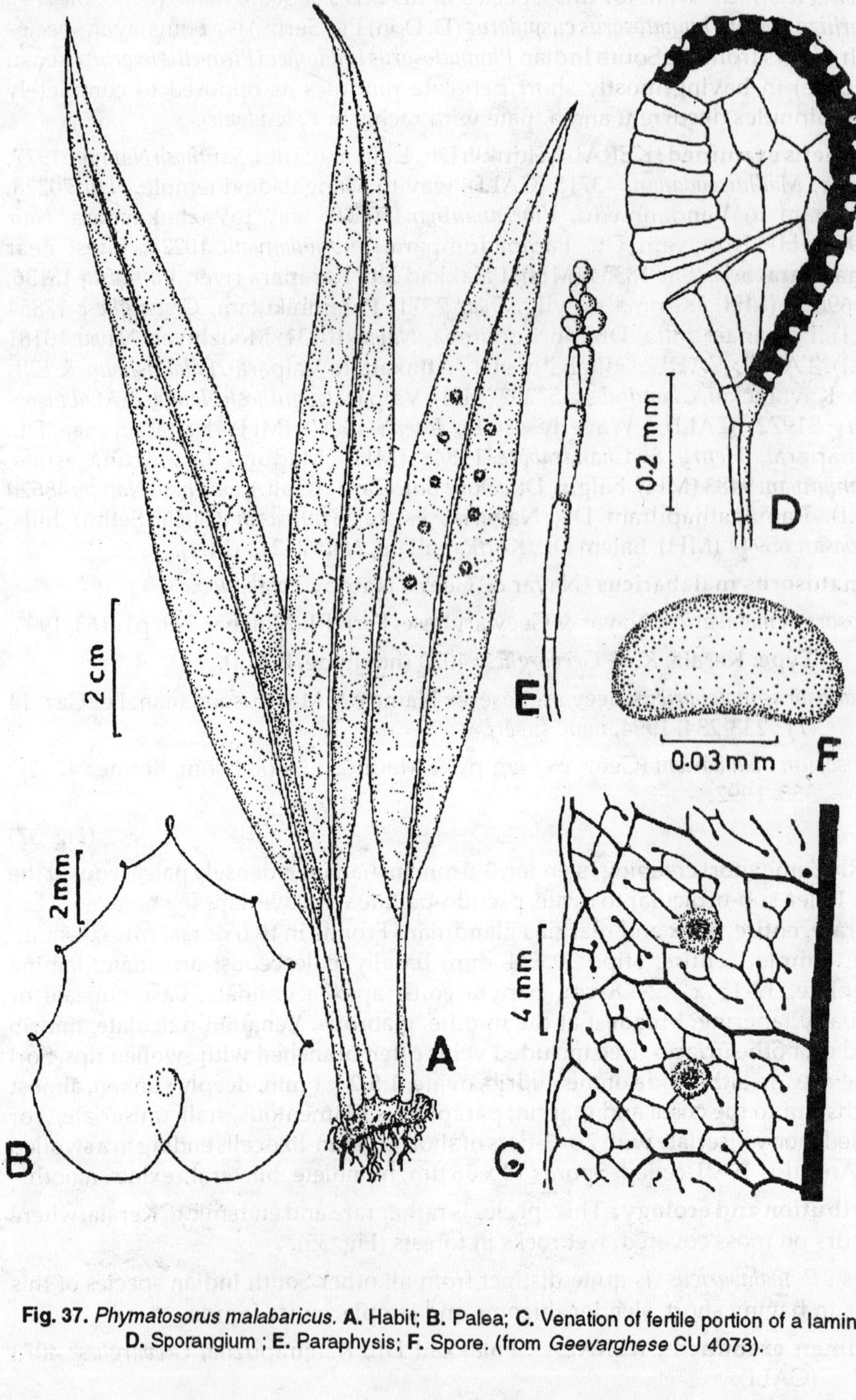

Fig. 37. *Phymatosorus malabaricus.* **A.** Habit; **B.** Palea; **C.** Venation of fertile portion of a lamina; **D.** Sporangium ; **E.** Paraphysis; **F.** Spore. (from *Geevarghese* CU 4073).

Phymatosorus nigrescens (Bl.) Pic. Serm., Webbia 28 : 459. 1973; Satija & Bir, Polyp. Ferns India 68. 1985; Manickam & Irudayaraj, Pterid. Fl. W. Ghats 317. pl. 242. 1992.

Polypodium nigrescens Blume, Enum. Pl. Jav. 2 : 126. 1828.

Pleopeltis longissima sensu Beddome, Ferns South India 59. t. 176. 1864, non Blume, 1828.

Phymatodes nigrescens (Bl.) J. Sm., Ferns Brit. & For. 94. 1866; Holttum, Fern Fl. Malaya 2 : 193.1954.

Pleopeltis nigrescens (Bl.) Beddome, Handb. Ferns Brit. India 367.1883.

Microsorum nigrescens (Bl.) Copeland, Occas. Papers Bishop Mus. 14 : 74. 1938; Sledge, Bull. Brit. Mus. (Nat. Hist.) Bot. 2 (5) : 144. 1960.

Microsorum alternifolium Copeland, Gen. Fil. 197. 1947, in part; Nayar, Bull. Nat. Bot. Gard. Lucknow 58: 12. 1961, non Willd., 1810.

(Figs. 14 g, h, 38, 40b)

Rhizome long creeping, 1-1.5 cm in diameter, fleshy, densely paleaceous at the apex and less so in older regions; dictyostelic, sclerenchyma strands numerous in the ground tissue. Palea ovate, 5 x 4 mm, peltate, clathrate, margins glandular. Fronds in two dorsal rows, 1.5 - 2 cm apart, pinnatifid; stipe cylindrical, 15-40 cm long, shiny, stramineous, articulated; phyllopodium 2-3 mm high; pinnules 6-12 with a broad wing (0.5 - 1 cm) on either side of the rachis, 20-24 x 2 -3 cm, glabrous, long-lanceolate, apex acuminate, entire, membranaceous, dark-green. Venation reticulate, free included veins simple or once forked with clavate apices. Sori in one row on either side of the midrib, round or oval, deeply sunken and protruding adaxially, 3 mm across; paraphyses with uniseriate stalk bearing series of balloon-shaped cells mostly in two rows. Annulus 19 or 20-celled; stalk 3-celled at the capsular base. Spores 40 x 25 µm, monolete, bilateral, exine smooth.

Distribution and ecology : *P. nigrescens* is distributed in North and South India, Sri Lanka, Malesia, Polynesia and Australia. It is fairly common in South India (Fig. 35) and occurs as a lowland epiphyte or lithophyte in moist, moderately shaded places especially near streams. The plant is often grown as an ornamental.

Notes : *P. nigrescens* is easily recognizable from other species of this genus by its membranaceous lamina, prominent veins and fronds with a few lateral pinnules (6-12 pairs) which are 2-3 cm broad. Nayar (1961b) observed the occurrence of uniseriate filamentous paraphyses in this species but the present study has shown that they have a 1-rowed stalk bearing two irregular rows of balloon-shaped cells.

Specimens examined : KERALA : Idukki Dt., Kaliyar (Cult.), *Santhosh Nampy* 51978; Thommenkuttu, *Santhosh Nampy* 48892 (CALI). Kollam Dt., Chalakkayam, *Mohanan* 59641 (MH). Kottayam Dt., Adukkam, Antony 1270 (MH). Kozhikode Dt., Anakampoil, *Santhosh Nampy* 48884 (CALI); Chedaleth, *Ellis* 19956 (MH); Jeerakapara, *Santhosh Nampy & Madhusoodanan* 45158; Vellarimala, *Geevarghese* 29466 (CALI). Palakkad Dt., Pandamala, *Geevarghese* 4053; Silent Valley, *Geevarghese* 4118 (CALI) , *Nair* 64367 (MH). Pathanamthitta Dt., Moozhiyar, *Nayar* 10085 (CALI). *Thiruvananthapuram* Dt., Kurisumalai, *Joseph* 46500; Ponmudi, *Nair & Ghosh*

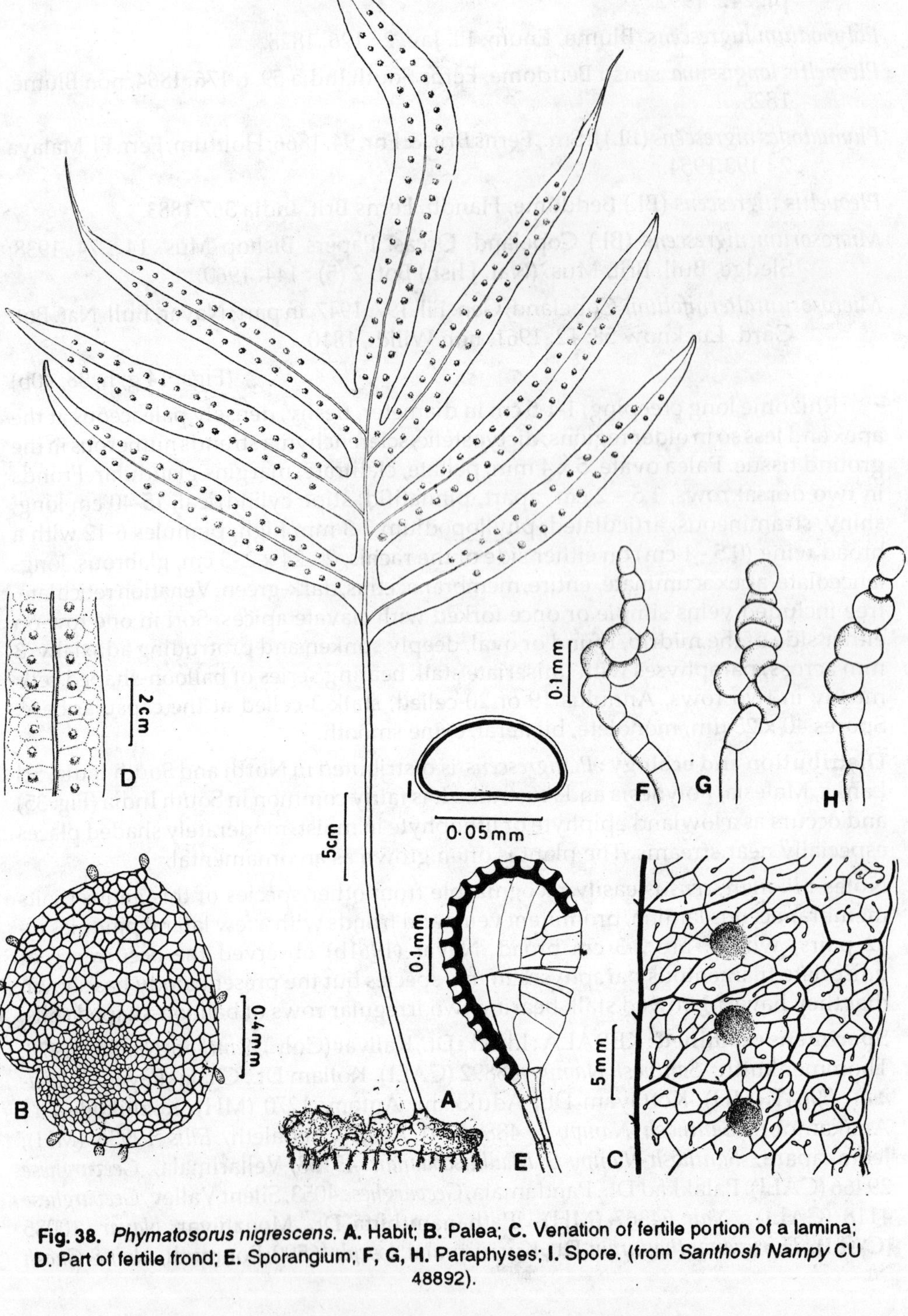

Fig. 38. *Phymatosorus nigrescens.* A. Habit; B. Palea; C. Venation of fertile portion of a lamina; D. Part of fertile frond; E. Sporangium; F, G, H. Paraphyses; I. Spore. (from *Santhosh Nampy* CU 48892).

51644; Suryakanthi river side, *Mohanan* 59437 (MH). Wynad Dt., Chandanathode, *Ellis* 25227 (MH); Periya, *Santhosh Nampy* 45958, *Santhosh Nampy & Madhusoodanan* 43766; Vythiri, *Geevarghese* 4173 (CALI).

Phymatosorus scolopendria (N.L. Burm.) Pic. Serm., Webbia 28 : 460. 1973; Satija & Bir, Polyp. Ferns India 70. 1985.

Polypodium scolopendria N. L. Burm., Fl. Ind. 232. 1768.

Polypodium phymatodes L., Mantl. Pl. Alt. 306. 1771.

Polypodium alternifolium Willd. *In* L., Sp. Pl. 4 : 168. 1810.

Pleopeltis phymatodes (L.) Beddome, Ferns South India 57. t. 173. 1864 & Handb. Ferns Brit. India 366. 1883.

Phymatodes scolopendria (N. L. Burm.) Ching, Contr. Inst. Bot. Nat. Acad. Peiping 2 : 63. 1933; Holttum, Fern Fl. Malaya 2 : 191. 1954.

Microsorum scolopendria (N.L. Burm.) Copeland, Univ.Calif. Publ. Bot. 16 : 112. 1929; Sledge, Bull. Brit. Mus. (Nat. Hist.) Bot. 2 (5) : 143. 1960; Nayar, Bull. Nat. Bot. Gard. Lucknow 58 : 10. 1961.

(Figs. 39, 40 d)

Rhizome long creeping, 6 mm in diameter, greyish-green, densely paleaceous at the apex and less so in older regions; dictyostelic, sclerenchyma strands numerous in the ground tissue. Palea ovate-lanceolate, peltate, dark-brown, clathrate, margins and apex glandular, dentate. Fronds in two dorsal rows, 2.2-5.5 cm apart, pinnatifid, fertile fronds longer than the sterile fronds; stipe 42 cm in fertile fronds and 25 cm in sterile fronds, purplish-green or purplish-brown, glabrous, articulated; pinnules 1-4 (or rarely more) pairs, lanceolate-oblong with a broad wing (3 cm) on the rachis, light or dark-green , glabrous, apex acute or acuminate, coriaceous. Midrib raised on both surfaces, other veins indistinct, free included veins branched with clavate apices. Sori usually in two (rarely more) irregular rows on either side of the midrib, circular to ovate, 1. 5 mm diameter, deeply sunken and forming protuberances adaxially; paraphyses absent. Sporangial stalk 3-celled at the capsular base, annulus 13 or 14-celled. Spores 27 x 19 µm, monolete, bilateral, light-yellow, exine smooth.

Distribution and ecology : *P. scolopendria* is distributed in Sri Lanka, India, China, Malesia to Polynesia, tropical Africa and Australia, where it occurs on exposed rocks and trees. This species has not so far encountered in the wild in South India, but had widely grown in gardens as a foliage plant (Fig. 43).

Notes : *P. scolopendria* is characterised by two (rarely more) irregular rows of sori on either side of the midrib, ovate-lanceolate, peltate palea with dentate margins, broadly winged rachis (2-5 cm) and pinnules broad (2-5 cm).

Specimens examined : KERALA : Idukki Dt., Kaliyar (Cult.), *Santhosh Nampy* 48628 (CALI). Kozhikode Dt., Azhchavattam (Cult.), *Madhusoodanan* 1181 ; Devagiri College Bot. Gard. (Cult.), *Santhosh Nampy* 51979 (CALI). Malappuram Dt., Calicut Univ. Bot. Gard. (Cult.), *Madhusoodanan* 29660 (CALI). Palakkad Dt., Govt. Victoria College Bot. Gard. (Cult.), *Sworrupanandan* 48629 (CALI). TAMIL NADU : Nilgiri Dt., *s. loc.* (Cult.?), *Beddome* 59995 (MH).

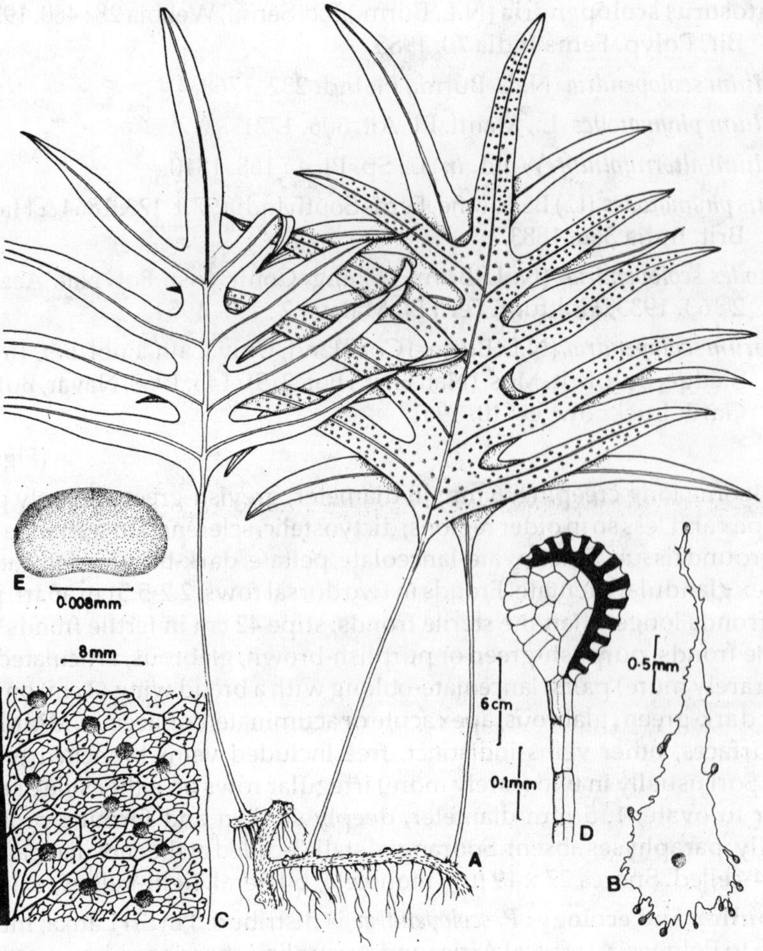

Fig. 39. *Phymatosorus scolopendria.* **A.** Habit; **B.** Palea; **C.** Venation of fertile portion of a lamina; **D.** Sporangium; **E.** Spore. (from *Santhosh Nampy* CU 51979).

PLEOPELTIS Humb. & Bonpl. ex Willd.

The name 'Pleopeltis' is derived from 'pleos', full; 'peltis', shield, the whole thus referring to the peltate paraphyses covering the immature sorus.

The genus as understood By Beddome (1883) was large and heterogenous. Many of the included species have since been segregated to other genera. Copeland

(1947) circumscribed it to include the Asiatic *Lepisorus* (J. Sm.) Ching and this was accepted by Sledge (1960). However , the present taxonomic consensus (Tryon & Tryon, 1982; Zink, 1988; Hennipman *et al.*, 1991) is in favour of keeping these two genera separate. Although *Pleopeltis* and *Lepisorus* share many characters in common, the former can readily be distinguished by its persistent laminar scales (additional notes are given under *Lepisorus).*

Pleopeltis with about 10 species is principally Neotropical. However, the distribution of two species, *P. macrocarpa* (Bory ex Willd.) Kaulf. and *P. macrosphaera* (Bak.) Panigr. & Patn. extend to South and South-east Asia.

Pleopeltis Humb. & Bonpl. ex Willd., Sp. Pl. 5 : 211. 1810; Beddome, Handb. Ferns Brit. India 344. 1883, in part; Copeland, Gen. Fil. 183. 1947, in part; Tryon & Tryon, Ferns Allied Plants 702. 1982; Hennipman *et al.* in Kubitzki (ed.), Fam. Gen. Vas. Plants 1 : 225. 1991.

Type species : *Pleopeltis angustata* Humb. & Bonpl.

Rhizome long creeping, slender, densely paleaceous all over; dictyostelic. Palea ovate-lanceolate, peltate, clathrate, margins dentate, glandular. Fronds simple, lanceolate, articulated; lamina entire, densely and prominently paleaceous abaxially, coriaceous. Veins anastomosing in a simple pattern, free included veins a few or none. Sori in one row on either side of the midrib, round or elliptic, borne at the union of several included veins; paraphyses umbrella-like, glandular, completely covering the sori in young stage. Annulus 13-celled. Spores monolete, bilateral, yellowish, shallowly to prominently verrucate.

Gametophyte : Under cultural conditions, the prothalli of *P. excavata* (Willd.) Sledge attain maturity in 3-5 months and of *P. normalis* in 10-12 months. The mature prothallus is cordate in the former whereas it is strap-shaped with a notched apex in the latter (Nayar, 1962).

Chromosome numbers : n = 34, 35, 37, *ca*. 70, 74; 2 n = *ca*. 210 (cf. Tryon & Tryon, 1982).

Distribution and ecology : *Pleopeltis* is distributed in America, Africa, Madagascar, Mascarenes to India and Sri Lanka. The genus is represented by a single species, *P. macrocarpa* in South India. It usually occurs on moist, moss covered rocks and tree trunks in shaded situations.

Pleopeltis macrocarpa (Bory ex Willd.) Kaulf., Berlin Jahrb. Pharm. 21 : 41. 1820; Sledge, Bull. Brit. Mus. (Nat. Hist.) Bot. 2 : 139. 1960; Bir & Trikha, Amer. Fern Jour. 58 : 123. 1968; Satija & Bir, Polyp. Ferns India 33. 1985; Manickam & Irudayaraj, Pterid. Fl. W. Ghats 330. pl. 251. 1992.

Polypodium macrocarpum Bory ex Willd., Sp. Pl. 5 : 147. 1810.

Type : Bourbon, *Bory s. n.* (isotype, P).

Polypodium lanceolatum L., Sp. Pl. 2 : 1082. 1753.

Polypodium marginale Bory ex Willd., Sp. Pl. 5 : 149. 1810, non L., 1753.

Polypodium lepidotum Willd. ex Schlecht., Adumbr. Pl. 2 : 17. t. 7. 1825.

Fig. 40. Habit of South Indian polypodioid ferns. **a.** *Phymatopteris montana*; **b.** *Phymatosorus nigrescens*; **c.** *Pleopeltis macrocarpa*; **d.** *Phymatosorus scolopendria*; **e.** *Pyrrosia lanceolata*; **f.** *P. ceylanica*; **g.** *P. porosa*.

Pleopeltis lepidota (Willd. ex Schlecht.) Presl, Tent. Pterid. 193. 1836; Beddome, Ferns South India 60. t. 181. 1864.

Pleopeltis lanceolata (L.) Presl, Tent. Pterid. 193. 1836, non Kaulf., 1824; Beddome, Handb. Ferns Brit. India 351. 1883.

Phlebodium lanceolatum (L.) J. Sm., Hook. Jour. Bot. 4 : 59. 1841.

Phlebodium macrocarpum (Bory ex Willd.) J. Sm., Hook. Jour. Bot. 4 : 59. 1841.

Drynaria macrocarpa (Bory ex Willd.) Fee, Gen. Fil. 270. 1852.

Drynaria lanceolata (L.) Fee, Gen. Fil. 270. 1852.

(Figs. 14 f, 40 c, 41)

Rhizome long creeping, 2-3 mm in diameter, dictyostelic, younger parts moderately paleaceous; sclerenchyma strands absent. Palea on the rhizome ovate-lanceolate and those on the fronds are either ovate-lanceolate or ovate-orbicular, 2.8 x 1 mm, peltate, centre dark-brown, margins pale, denticulate, glandular at apex. Fronds simple, 12 x 1.3 cm, in two regular rows, 1-2 cm apart, moderately scaly abaxially; stipe 1-2 x 0.1 cm, articulated, sparsely paleaceous; lamina lanceolate, entire or wavy, coriaceous, acute at apex, base decurrent. Veins indistinct externally, areoles obliquely elongate, free included veins occasionally present. Sori distinct, one row on either side of the midrib, 3 mm in diameter, ovate or elliptic, median, on the anterior 2/3 of the lamina, completely covered in young stage by peltate, umbrella- shaped paraphyses. Annulus 13-celled, stalk 3-celled at the capsular base. Spores 68 x 49 µm, monolete, bilateral, yellowish, exine verrucate.

Distribution and ecology : *P. macrocarpa* is distributed in South India, Sri Lanka, tropical America, tropical and southern Africa including St. Helena, Madagascar and Mascarenes. It occurs as an epiphyte on moss covered trees in the higher mountains of South India (Fig. 43).

Notes : Linnaeus (1753) originally included this species under the genus *Polypodium* as *P. lanceolatum* L. Subsequently, it has been transferred to the genus *Pleopeltis* by Presl in 1836. The epithet 'lanceolatum' of Linnaeus cannot be maintained under *Pleopeltis* as it has already pre-occupied in the genus *Pleopeltis* based on a different type. The next earliest available epithet 'macrocarpum' (Willd.) Kaulf. is adopted for the present species.

According to Pichi Sermolli (1965) *P. macrocarpa* is polymorphic with respect to variability in the shape and size of the frond, the dimensions of the sori, the quantity of the paleae on the surface of the lamina and the degree of thickness of wall of the cells of palea. However, the South Indian specimens belonging to this species are rather uniform in having simple, lanceolate fronds of about 12 x 1.3 cm, their sori measuring about 3 mm, the lamina densely paleaceous abaxially and having the palea with dark sclerotic centre.

Specimens examined : KERALA : Thiruvananthapuram Dt., Ponmudi, *Santhosh Nampy* 48879 (CALI). TAMIL NADU : Nilgiris Dt., Kodanad R. F., *Vajravelu* 36804; Long Wood R. F. (Kotagiri), *Vajravelu* 35137 (MH), *Santhosh Nampy* 48625;

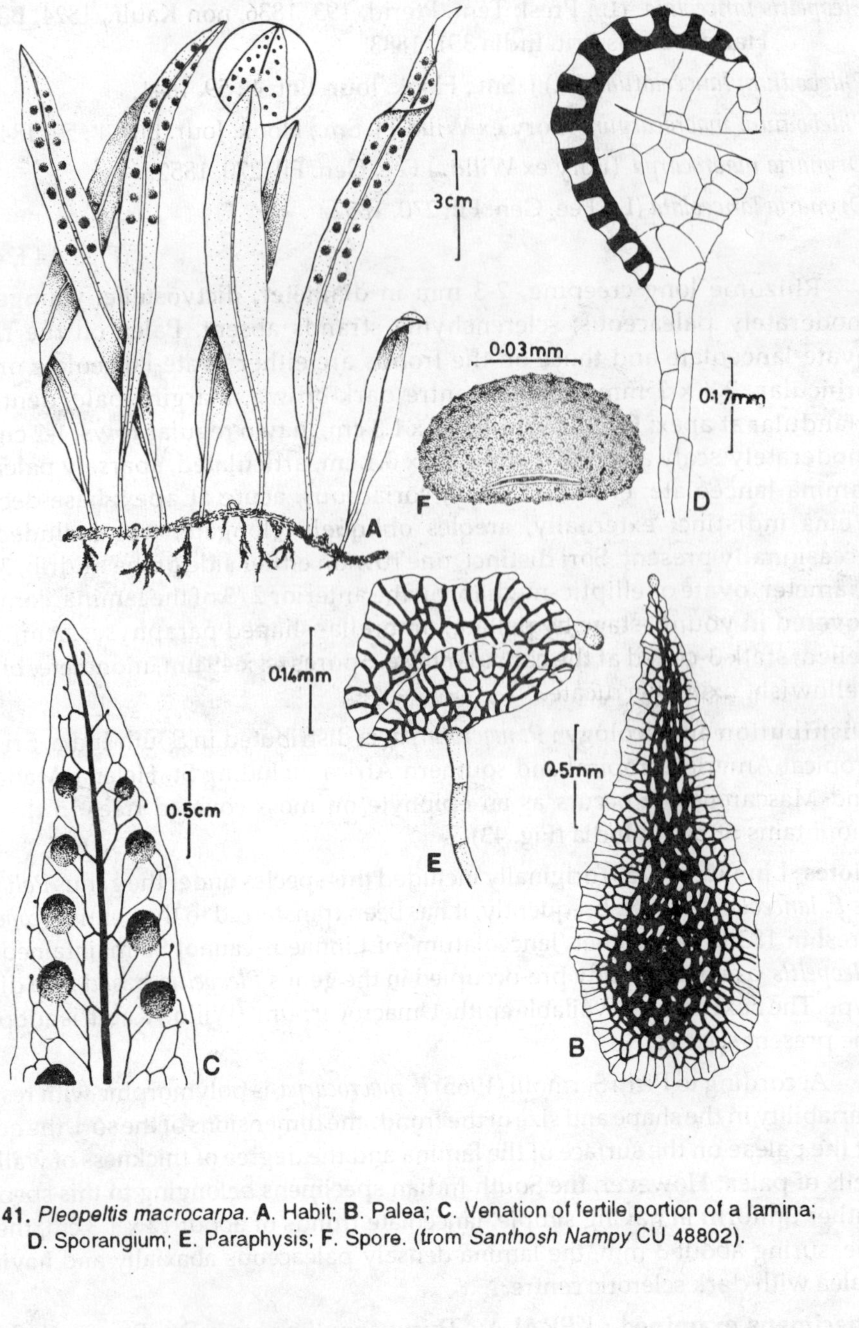

Fig. 41. *Pleopeltis macrocarpa*. **A.** Habit; **B.** Palea; **C.** Venation of fertile portion of a lamina; **D.** Sporangium; **E.** Paraphysis; **F.** Spore. (from *Santhosh Nampy* CU 48802).

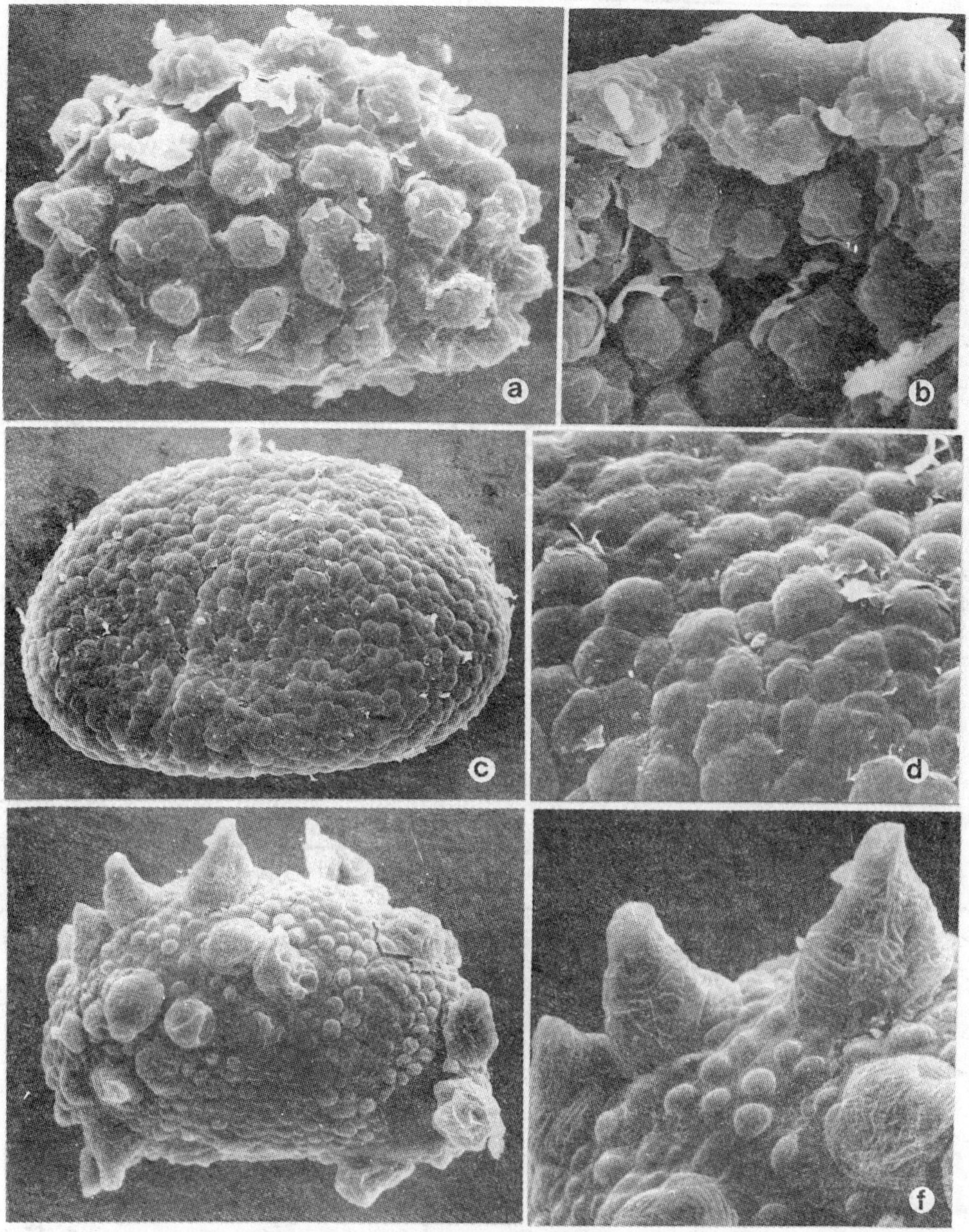

Fig. 42. SEM photograph of spores of polypodioid ferns. **a, b.** *Phlebodium aureum (John & Stone* 25541); **c, d.** *Pleopeltis macrocarpa (Manickam* 960); **e, f.** *Pyrrosia heterophylla (Davidse* 8102). **a, c, e.** Lateral view; **b, d, f.** A part enlarged. **a.** x 1500; **c, e.** x 1125; **b, f.** x 2625; **d.** x 3750.

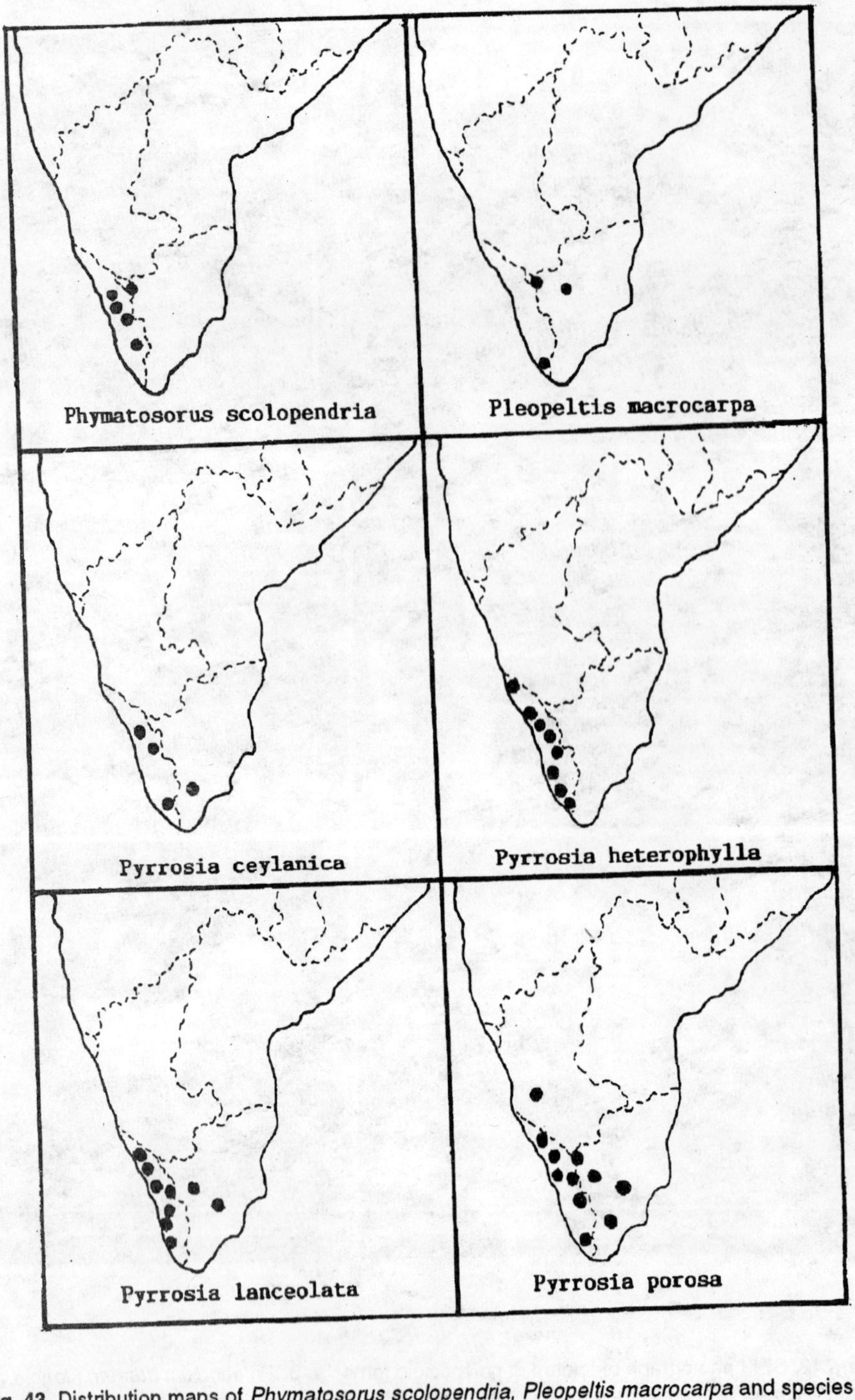

Fig. 43. Distribution maps of *Phymatosorus scolopendria, Pleopeltis macrocarpa* and species of *Pyrrosia* in South India.

Naduvattam, *Santhosh Nampy* 45160; Ooty. Bot. Gard. (wild), *Santhosh Nampy* 48802 (CALI). Salem Dt., Yercaud (Badmadies estate), *Subramanyam* 7553 (MH).

PYRROSIA Mirbel

The nomenclature of the genus *Pyrrosia* has been a matter of discord since long. The genus was originally established by Mirbel (1803) to include a single species, *P. chinensis* Mirbel. In the same work he also described another closely allied genus *Candollea* to include four species, but did not state that any one of the four species he included is "typical" or otherwise to be regarded as type. *Pyrrosia* being monotypic, the included species (*P. chinensis)* will stand as its type.

Desvaux (1811, 1827) accepted Mirbel's genera, but renamed *Candollea* as *Cyclophorus* as the former name was used by Labillardier (1806) for another genus. Meanwhile, Kaulfuss (1824) renamed *Cyclophorus* Desv. as *Niphobolus* to avoid confusion as the former name, has already been attributed for another taxon. This name, although antedated Mirbel's and Desvaux's genera, was adopted by several authors (Blume, 1828, 1829; Presl, 1836; J. Smith, 1842, 1875; Hooker, 1842; Fee, 1850-52; Beddome, 1864, 1883; Giesenhagen, 1901) until Farwell (1931) recognised that *'Pyrrhosia'* is the correct name of the genus. Deviating from the nineteenth century botanists, Underwood (1903) and Christensen (1906) adopted the name *Cyclophorus* for *Pyrrosia*. The matter has already been discussed at length by Pichi Sermolli (1953), Shing (1983) and Hovenkamp (1986). However, the present taxonomic opinion is in favour of uniting Mirbel's genera *Pyrrosia* and *Candollea* under the name *Pyrrosia* (Ching, 1935; Pichi Sermolli, 1953; Nayar, 1957, 1961 c; Nayar & Chandra, 1965, 1967; Nayar & Kaur, 1974; Shing, 1983; Satija *et al.*, 1983; Satija & Bir, 1985; Hovenkamp, 1984, 1986; Manickam, 1986; Chandra & Kaur, 1987; Hennipman *et al.*, 1991; Manickam & Irudayaraj, 1992; Nampy & Madhusoodanan, 1992; Nayar & Geevarghese, 1993).

The earliest extensive work on the genus was that of Giesenhagen (1901) who treated fifty species of *Pyrrosia* under the name *Niphobolus* Kaulf. The genus *Pyrrosia* was recently monographed by Hovenkamp (1986) and recognised fifty one species which include species formerly treated under genera such as *Drymoglossum* Presl, *Saxiglossum* Ching, *Pteropsis* Desv. and *Neoniphopsis* Nakai.

Pyrrosia differs from other genera of Polypodiaceae in having thick, fleshy fronds with dense stellate tomentum, compyloneuroid venation and the stoma being completely encircled by an epidermal cell.

Pyrrosia Mirbel, Hist. Nat. Gen. 4 : 70. 1803; Farwell, Amer. Midl. Nat. 12 : 245. 1931, as *Pyrrhosia;* Ching, Bull. Chin. Bot. Soc. 1 : 36. 1935; Copeland, Gen. Fil. 192. 1947; Holttum, Biol. Rev. 24 : 282. 1949; Pic. Serm., Webbia 9 : 441. 1953; Holttum, Fern Fl. Malaya 2 : 141. 1954; Nayar, Jour. Ind. Bot. Soc. 40 (2) : 164. 1961; Nayar & Chandra, Bull. Nat. Bot. Gard. Lucknow 117. 1965 & Can. Jour. Bot. 45 : 615. 1967; Devol & Kuo in Li *et al.* (eds.), Fl. Taiwan 1 : 205. 1975; Satija *et al.*, Bull. Bot. Surv. India 25 (1-4) : 62-89. 1983; Shing,

Amer. Fern Jour. 73 (3): 73-77. 1983; Hovenkamp, Leiden Bot. Ser. 9 : 1-260. 1986; Hennipman *et al.* in Kubitzki (ed.), Fam. Gen. Vas. Plants 211. 1991; Manickam & Irudayaraj, Pterid. Fl. W. Ghats 321. 1992; Nayar & Geevarghese, Fern Fl. Malabar 406. 1993.

Type species : *Pyrrosia chinensis* Mirbel [= *Pyrrosia stigmosa* (Sw.) Ching].

Candollea Mirbel, Hist. Nat. Gen. 69. 1803, non *Candollea* Labell., 1806.

Cyclophorus Desv., Berlin Mag. 5 : 300. 1811; Underwood, Bull. Torr. Bot. Club 30 : 670. 1903.

Niphobolus Kaulf., Enum.. Fil. 124. 1824, *nom. superfl.*; Beddome, Ferns South India 61. 1864 & Handb. Ferns Brit. India 324. 1883; Giesenh., Niphobolus 1901.

Pteropsis Desv., Mem. Soc. Linn. Par. 6 : 218. 1827, *nom. rej.*

Drymoglossum Presl, Tent. Pterid. 227. 1836; Nayar, Jour. Ind. Bot. Soc. 36 (2): 169. 1957.

Galeoglossa Presl, Epim. Bot. 132. 1851.

Scytopteris Presl, Epim. Bot. 133. 1851.

Sphaerostichum Presl, Epim. Bot. 134. 1851.

Polycampium Presl, Epim. Bot. 135. 1851.

Apalophlebia Presl, Epim. Bot. 137. 1851.

Gyrosorium Presl, Epim. Bot. 139. 1851.

Niphopsis J. Sm., Cat. Cult. Ferns 6. 1857.

Neoniphopsis Nakai, Bot. Mag. Tokyo 42 : 21. 1928.

Saxiglossum Ching, Contr. Inst. Bot. Nat. Acad. Peiping 2 : 1-4. 1933.

Rhizome short or long creeping, slender or thick, densely paleaceous; dictyostelic, sclerenchyma strands present or absent. Palea lanceolate or shield-like, basifixed or pseudo-peltate or peltate, margins smooth or ciliate. Fronds monomorphic or dimorphic, articulated, sessile or stipitate; lamina entire or rarely dissected, fleshy, coriaceous, usually densely set with stellate indumentum abaxially, sparsely so or glabrous adaxially; indumentum mono- or dimorphic. Veins hidden, compyloneuroid, included and excluded veins simple, free. Sori round, terminal on the included veins, superficial to deeply sunken, densely covered by stellate indumentum of the lamina. Sporangia stalked, annulus usually 15 to 22-celled. Spores monolete, bilateral, yellowish, exine tuberculate to verrucate.

Gametophyte : The mature prothallus of *Pyrrosia* is cordate with a thin midrib and expanded wings (Nayar & Chandra, 1965).

Chromosome numbers : n = 36, 37, 74, 108-111; 2n = 72, ± 73, 74, ± 144, ± 149, ± 179, 216 (cf. Hennipman *et al.*, 1991).

Distribution and ecology : *Pyrrosia* is an almost exclusively paleotropical genus distributed mainly in Asia and a few species in Australia, New Zealand, and Africa. Most of its species are epiphytes on exposed trees in lowlands or mountain forests or terrestrial at low to higher elevations. A few species also grow on mossy rocks at medium to higher altitudes.

Key to the species

1a. Rhizome long creeping 2
1b. Rhizome short creeping .... *P. porosa*
2a. Sori distinct 3
2b. Sori acrostichoid .... *P. heterophylla*
3a. Palea entire .... *P. ceylanica*
3b. Palea ciliate .... *P. lanceolata*

Pyrrosia ceylanica (Giesenh.) Sledge, Bull. Brit. Mus. (Nat. Hist.) Bot. 2 (5) : 133. 1960; Abeywickrama, Ceyl. Jour. Sci. Sect. A 13 : 26. 1956, invalid comb. ; Satija *et al.*, Bull. Bot. Surv. India 25 (1-4) : 88. 1983, as doubtful record; Bhargavan & Nair, Jour. Econ. Tax. Bot. 6 (2) : 269. 1985; Satija & Bir, Polyp. Ferns India 53. 1985, as doubtful record; Hovenkamp, Leiden Bot. Ser. 9 : 163. 1986.

Niphobolus ceylanicus Giesenh., Niphobolus 216. f. 19. 1901.

Type : Sri Lanka, *G. Wallich s. n.,* (holotype, P; isotype, M. fragments).

Cyclophorus ceylanicus C. Chr., Ind. Fil. 198. 1906.

Pyrrosia lanceolata sensu Manickam & Irudayaraj, Pterid. Fl. W.Ghats 322. 1992, in part, non *Acrostichum lanceolatum* L., 1753.

(Figs. 40f, 44)

Rhizome long creeping, slender, 0.7 - 1 mm in diameter; cortical sclerenchyma sheath prominent, sclerenchyma strands a few or usually one situated centrally in the ground tissue, dictyostelic, vascular bundles 3. Palea more dense towards the apex and less so in older regions of the rhizome, linear-lanceolate, 4 x 0.6 mm, shining light-brown, peltate, base trilobed or slightly lobed, margins entire, apex simple or occasionally once forked, non-glandular. Fronds simple, moderately dimorphic, 0.9- 4.5 cm apart, articulated; fertile fronds : stipe 1-28 x 6 mm; lamina 25-74 x 4-7 mm, base cuneate, widest at or below the middle, apex acute to obtuse, margins entire, thick, fleshy, densely stellate hairy abaxially, glabrous adaxially; sterile fronds: stipe 4 - 25 mm; lamina 16-40 x 4 -11 mm, otherwise similar to the fertile ones. Hairs monomorphic, persistent, usually 9-11 rayed, rays broad. Venation compyloneuroid, veins hidden, midrib not raised on either surface, grooved adaxially, lateral veins distinct, secondary veins form regular areoles with simple free excurrent included veins. Sori apical or almost completely covering the under surface, sunken in the leaf tissue, closely packed; individually distinct, at the end of included veins, exserted from the indument; paraphyses stellate, in a central bundle in the sorus, similar to the laminar hairs. Sporangia stalked, stalk 1/2 as long as the capsule, annulus 17 or 18-celled. Spores 71 x 46 µm, monolete, bilateral, yellowish, exine strongly verrucate.

Distribution and ecology : *Pyrrosia ceylanica* is fairly common in South India (Fig. 43), and is also distributed in Sri Lanka. It is an epiphyte in wet, evergreen forests of 700- 750 m altitude.

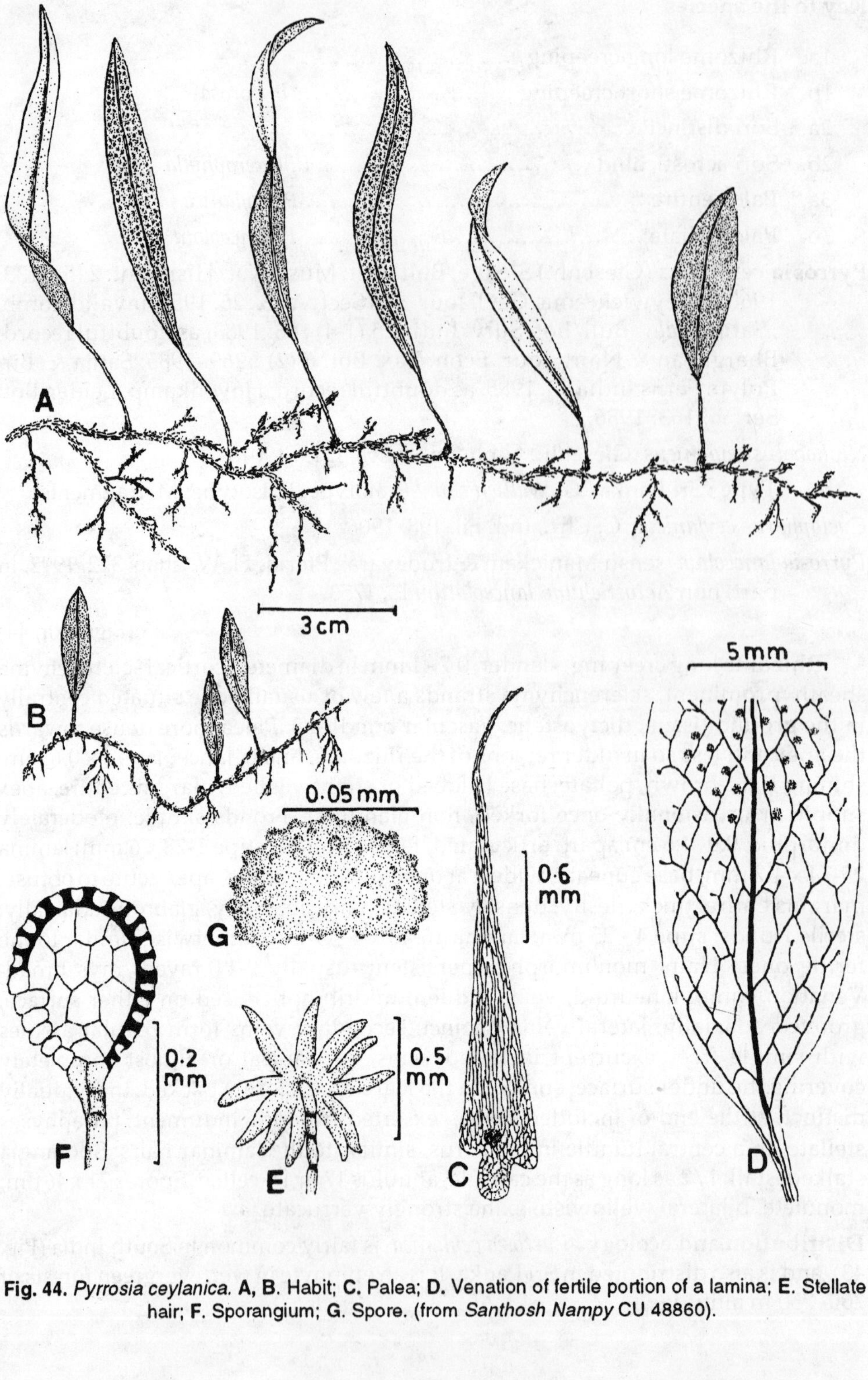

Fig. 44. *Pyrrosia ceylanica.* A, B. Habit; C. Palea; D. Venation of fertile portion of a lamina; E. Stellate hair; F. Sporangium; G. Spore. (from *Santhosh Nampy* CU 48860).

Notes : Giesenhagen (1901) originally treated this species under the genus *Niphobolus.* Subsequently, Abeywickrama *(Ceyl. Jour. Sci. Sect.* A 13 : 26. 1956) transferred it to the genus *Pyrrosia,* but he failed to designate the basionym of his new combination. However, Sledge (1960) validated the name *Pyrrosia ceylanica* referring to its basionym.

This species has been frequently confused with *P. lanceolata* (L.) Farwell, but can be distinguished by its glabrous palea as opposed to marginally ciliate palea in *P. lanceolata.* *P. ceylanica* can be distinguished from other species of this genus by its wiry rhizome with shining light brown palea and somewhat shorter fronds.

Specimens examined : KERALA : Kozhikode Dt., Anakampoil, *Santhosh Nampy* 48860; way to Olichuchattam, *Santhosh Nampy* 48862 (CALI). Palakkad Dt., Nelliampathy, *Geevarghese* 17810; Panthenthode, *Nayar* 10601 ; Silent Valley, *Geevarghese* 4056 (CALI). Pathanamthitta Dt., Karimala, *Nayar* 7379 (CALI). Wynad Dt., Periya, *Santhosh Nampy* 45965 (CALI). TAMIL NADU : Kattabomman Dt., Kothayar, *Madhusoodanan* 44859, *Santhosh Nampy* 48631 (CALI).

Pyrrosia heterophylla (L.) Price, Kalikasan 3 : 177. 1973; Hovenkamp, Leiden Bot. Ser. 9 : 299. 1986; Chandra & Kaur, Nomen. Guide Bedd. Ferns South India & Ferns Brit. India 22. 1987.

Acrostichum heterophyllum L., Sp. Pl. 2 : 1067. 1753.

Type : Sri Lanka, *Paul Hermann s. n.* (holotype, BM).

Niphobolus nummularifolius auct. non, (Sw.) J. Sm.: Beddome's Ferns South India 62. t. 186. 1864, in part.

Drymoglossum beddomei Clarke, Trans. Linn. Soc. Lond. 2 : 575. 1880, *nom. nud.*

Drymoglossum piloselloides var. *beddomei* Beddome, Handb. Ferns Brit. India 413. 1883, *nom. invalid.*, art. 34 ICBN.

Drymoglossum heterophyllum (L.) Trimen, Jour. Linn. Soc. 24 : 152. 1887; C. Chr., Ind. Fil. 246. 1906, excl. syn. in part; Sledge, Bull. Brit. Mus. (Nat. Hist.) Bot. 2 (5) : 135. 1960; Satija & Bir, Polyp. Ferns India 54. 1985; Manickam, Fern Fl. Palni Hills 148. 1986; Manickam & Irudayaraj, Pterid. Fl. W. Ghats 310. pl. 237. 1992; Nayar & Geevarghese, Fern Fl. Malabar 380. 1993.

Drymoglossum heterophyllum var. *beddomei* C. Chr., Dansk. Bot. Ark. 6 : 85. pl. 128. 2 a-c. 1929.

Pyrrosia piloselloides sensu Nampy & Madhusoodanan, Indian Fern Jour. 9 : 215. 1992, non L., 1753.

Drymoglossum piloselloides sensu Nayar & Geevarghese, Fern Fl. Malabar 381. 1993, non L., 1753.

-Maratte-mala-maravara Rheede, Hort. Malab. 12 : 57. t. 29. 1693.

(Fig. 45)

Rhizome long creeping, slender, 0.7 mm in diameter, densely paleaceous all over; cortical sclerenchyma sheath present, sclerenchyma strands absent in the ground tissue, dictyostelic, vascular bundles 3 or 4. Palea ovate-lanceolate to ovoid, 1-2 x 0.5 mm, peltate, uniformly pale-brown except the attaching region where it is dark-brown, margins dentate to ciliate. Fronds dimorphic, simple, stipitate,

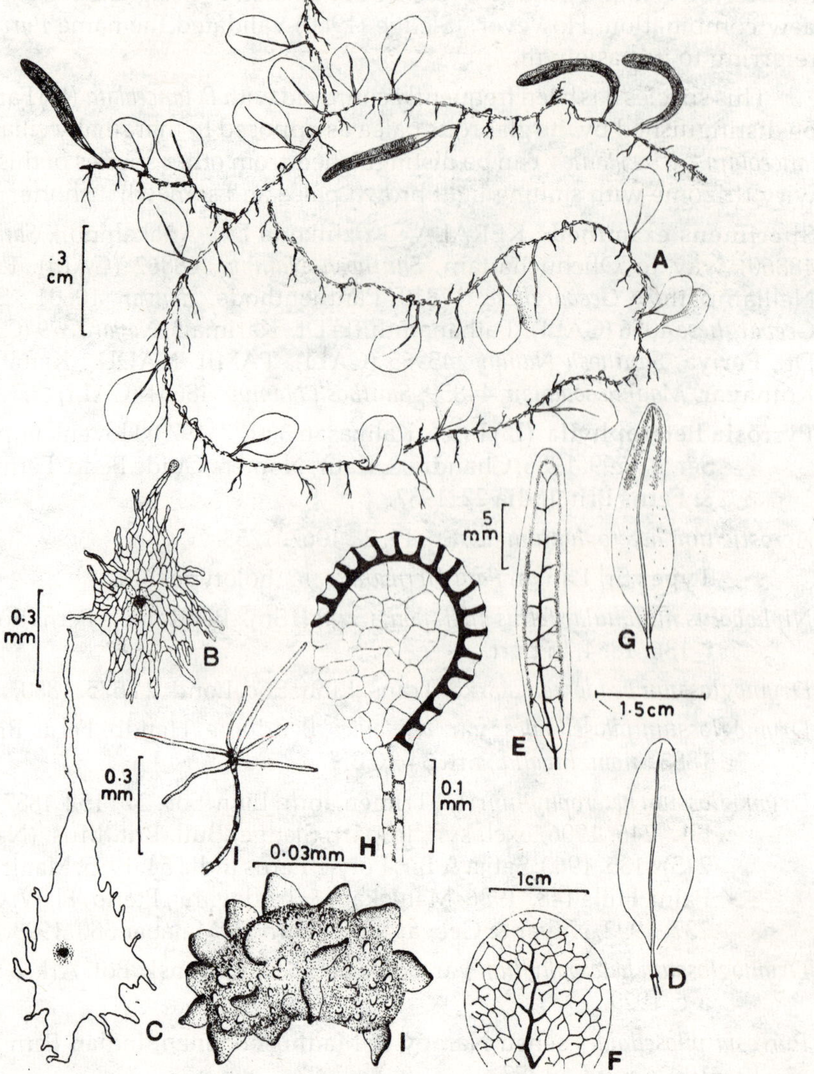

Fig. 45. *Pyrrosia heterophylla*. **A.** Habit; **B, C.** Palea; **D.** Sterile frond; **E.** Venation of fertile frond; **F.** Venation of sterile frond; **G.** Fertile frond; **H.** Sporangium ; **I.** Paraphysis; **J.** Spore. (from *Santhosh Nampy* CU 48633).

articulated, distantly placed, *ca.* 1.5 cm apart, stipe 1-4 mm; sterile fronds : orbicular or ovate or elliptic or lanceolate, widest at the middle, thick, fleshy, entire, apex rounded, 1.5 - 2 x 1.5 cm, glabrous when mature; fertile fronds : oblong-lanceolate to linear-lanceolate, 25-50 x 4-7 mm, apex rounded. Midrib of fertile fronds raised abaxially, veins hidden, anastomosing to form larger costal areoles and smaller marginal areoles, free included veins simple or forked, recurrent. Indument scattered on the fronds, stellate, rays 6 or 7, sinuous. Sori linear, all along the undersurface of the lamina except the midrib, continuous; receptacle along the outermost connecting vein, submarginal between the costa and margin; mixed with stellate hairs. Sporangial stalk 1.5 - 2 times longer than the capsule, annulus 18-celled. Spores 75 x 48 µm, monolete, bilateral, warty with conical spines.

Distribution and ecology : *Pyrrosia heterophylla* is distributed in South India (Fig. 43), Sri Lanka and Seychelles. It grows on exposed trees and rocks in low to medium elevations.

Notes : In 1864, Beddome illustrated a specimen in his *Ferns of South India* (pl. 186) under the name *Niphobolus nummularifolius* based on a specimen from Anamalais. This is characterised by stellate indumentum on the fronds, sharp pointed rhizome scales and sori generally covering the whole back of the linear fronds. Subsequently, he (Beddome, 1866) figured yet another closely related taxon under the same name in his *Ferns of British India* (pl. 320) based on a specimen from Assam. This specimen is characterised by ferruginous tomentose fronds, bright ferruginous fibrillose rhizome scales and coenosori covering the whole surface of the lamina. But he added a note that the plate 186 of the *Ferns of South India* figured by him under the name *Niphobolus nummularifolius* is only a form of *Drymoglossum piloselloides.* We have examined a large number of specimens belonging to this species complex both from South India and from Assam and is of the opinion that the South Indian collections clearly pertain to *Pyrrosia heterophylla* (= *D. heterophyllum*). It differs from *P. piloselloides* in having acrostichoid fertile fronds as against sori restricted to a marginal band running from the apex to the almost base in *P. piloselloides.* This character is very evident in all specimens from Assam. We have not encountered any specimen of *P. nummularifolia* and *P. piloselloides* from South India. This is in corroboration with Hovenkamp (1986). However, Nayar and Geevarghese (1993) seem to have confused in recognising *Drymoglossum heterophyllum* from *D. piloselloides* in South India. They state that the latter is "common throughout the plains of Malabar", which is contrary to the present observations.

The plant is commonly known in Malayalam as "Seethathali" and is locally used as a cooling agent, for the treatment of swellings, sprains and for relieving pain (Nayar, 1959).

Specimens examined : KERALA : Alappuzha Dt., Cherthala, *Santhosh Nampy & Madhusoodanan* 45175 (CALI). Idukki Dt., Kaliyar, *Santhosh Nampy* 48634 (CALI). Kannur Dt., Kannavam R. F., *Santhosh Nampy* 12851 (CALI). Kollam Dt., Kulathupuzha, *Santhosh Nampy* 48633, *Madhusoodanan* 48636 (CALI). Kozhikode Dt., Peruvannamuzhi, *Geevarghese* 29457 (CALI). Malappuram Dt., Calicut Univ.

campus, *Sivarajan* 1113, *Uma Maheswari* 338; Thiruvangad, *Santhosh Nampy* 48635, *Madhusoodanan* 21293 (CALI). Palakkad Dt., Kaikatty, *Nayar* 7306; Karean shola, *Nayar* 7375; Noorani, *Geevarghese* 17814; Pandikadu, *Madhusoodanan* 21346, *Nayar & Madhusoodanan* 21386; Pothundi, *Omana* 18650 (CALI). Thiruvananthapuram Dt., way to Agasthiamala, *Santhosh Nampy & Madhusoodanan* 48637 (CALI).

Pyrrosia lanceolata (L.) Farwell, Amer. Midl. Nat. 12 : 245. 1930, as *Pyrrhosia lanceolatus*; Ching, Bull. Chin. Bot. Soc. 1 : 70. 1935 ; Sledge, Bull. Brit. Mus. (Nat. Hist.) Bot. 2 (5) : 133. 1960; Nayar, Jour. Ind. Bot. Soc. 40 : 164-186. 1961; Nayar & Chandra, Bull. Nat. Bot. Gard. Lucknow 117 : 58. 1965 & Can. Jour. Bot. 45 : 615- 634. 1967; Nayar & Kaur, Comp. Bedd. Handb. Ferns Brit. India 80. 1974; Satija *et al.*, Bull. Bot. Surv. India 69. 1983; Satija & Bir, Polyp. Ferns India 38. 1985; Manickam, Fern Fl. Palni Hills 150. 1986; Manickam & Irudayaraj, Pterid. Fl. W. Ghats 322. 1992, in part.

Acrostichum lanceolatum L., Sp. Pl. 2 : 1067. 1753; Burm., Fl. Ind. 228. 1768; Ching, Bull. Chin. Bot. Soc. 1 : 46. 1935.

Type : Sri Lanka, *Paul Hermann* I, fol. 3 (holotype, BM; isotype, L).

Polypodium adnascens Sw., Syn. Fil. 25, 222. p1. 2. fig. 2. 1806.

Polypodium spissum Bory ex Willd., Sp. Pl. 5 : 146. 1810; Ching, Bull. Chin. Bot. Soc. 1 : 46. 1935.

Cyclophorus adnascens (Sw.) Desv., Mag. Ges. Naturf. Freunde Berlin 5 : 300. 1811.

Niphobolus adnascens (Sw.) Kaulf., Enum. Fil. 124. 1824; Blume, Enum. Pl. Jav. 2 : 275. 1828; Beddome, Ferns South India 62. pl. 184. 1864 & Handb. Ferns Brit. India 325. t. 176. 1883; Giesenh., Niphobolus 196, 206, 209. t. 18. 1901.

Niphobolus varius Kaulf. , Enum. Fil. 125. 1824; Giesenh., Niphobolus 151, 198, 208. 1901.

Gymnopteris lanceolata Moore, Ind. Fil. 10. 1857.

Niphobolus nudus Giesenh., Niphobolus 149. 1901.

Cyclophorus nudus (Giesenh.) C. Chr., Ind. Fil. 200. 1906.

Pyrrosia varia (Kaulf.) Farwell, Amer. Midl. Nat. 12 : 302. 1931; Ching, Bull. Chin. Bot. Soc. 1 : 70. 1935; Holttum, Fern Fl. Malaya 2 : 146. t. 61. 1954; Nayar, Jour. Ind. Bot. Soc. 40 : 164-186. 1961; Nayar & Chandra, Bull. Nat. Bot. Gard. Lucknow 117 : 87. 1965 & Can. Jour. Bot. 45 : 615-634. 1967; Nayar & Kaur, Comp. Bedd. Handb. Ferns Brit. India 80. 1974.

Pyrrosia adnascens (Sw.) Ching, Bull. Chin. Bot. Soc. 1 : 45. 1935; Holttum, Fern Fl. Malaya 2 : 144. t. 60. 1954; Nayar, Jour. Ind. Bot. Soc. 40 : 164-186. 1961; Nayar & Chandra, Bull. Nat. Bot. Gard. Lucknow 117 : 48. 1965 & Can. Jour. Bot. 45: 615-634. 1967; Nayar & Kaur, Comp. Bedd. Handb. Ferns Brit. India 80. 1974; Devol & Kuo in Li *et al.* (eds.), Fl. Taiwan 1 : 205. 1975.

Pyrrosia caudata Ching, Bull. Chin. Bot. Soc. 1 : 46. 1935.

Pyrrosia nuda (Giesenh.) Ching, Bull. Chin. Bot. Soc. 1 : 70. 1935.
(for more synonyms see Hovenkamp, 191-194. 1986).

- Tiri-panna Rheede, Hort. Malab. 12 : 141. t. 74. 1693.

(Figs. 14 i, 40 e, 46)

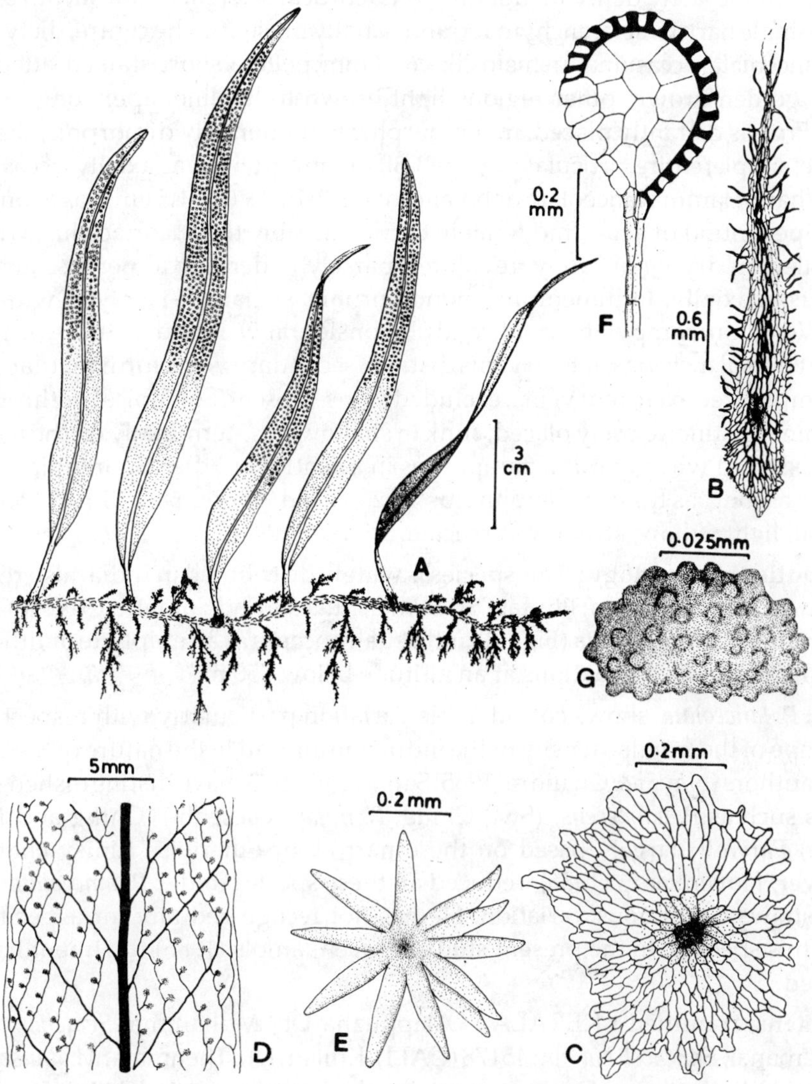

Fig. 46. *Pyrrosia lanceolata.* **A.** Habit; **B, C.** Palea ; **D.** Venation of fertile portion of a lamina; **E.** Stellate hair; **F.** Sporangium ; **G.** Spore. (from *Santhosh Nampy* CU 44841)

Rhizome long creeping , slender, 1-1.5 mm in diameter, densely paleaceous at the apex, moderately dense in older regions; cortical sclerenchyma sheath 3-4 cells thick, a single narrow sclerenchyma strand usually present at the centre, dictyostelic. Palea lanceolate, occasionally shield-like, 6 x 1 mm, peltate, short-stalked, attachment region golden-brown, other regions light-brown to hyaline, apex long, margins ciliate. Fronds distantly placed, monomorphic to moderately dimorphic, distinctly stipitate, simple, entire, articulated; stipe 3-30 x 1 mm, grooved adaxially; occasionally stellate hairy; lamina lanceolate or oblanceolate, 2-16.4 x 0.4 -1.5 cm, base cuneate or acute, apex round or acute, moderately to considerably thick, coriaceous, wrinkled or inrolled on drying or rarely revolute, sparsely to densely tomentose abaxially, glabrous adaxially. Indumentum monomorphic, stellate, 7-11 rayed with broad arms. Venation compyloneuroid, midrib considerably raised or not so, with or without dorsal incisions, lateral veins distinct, secondary veins form regular areoles with simple, free excurrent veins, excluded veins present. Sori apical or throughout the lamina, distinct, closely placed, sunk in shallow pits, terminally on the included veins, exserted when mature; paraphyses in a central bundle. Sporangia stalked ; stalk 1 x as long as the capsule; annulus 16-20-celled. Spores 68 x 48 µm, monolete, bilateral, light yellow, strongly verrucate.

Distribution and ecology: This species is widely distributed in India, also reported to occur in China, Taiwan, Bhutan, Sri Lanka, Malesia and Sumatra to New Guinea. It is common in South India (Fig. 43) and usually occurs as an epiphyte or lithophyte generally in exposed situations at an altitude below 750 m.

Notes : *P. lanceolata* shows considerable variation particularly with respect to size and shape of the fronds, density of the indumentum and in the nature of the midrib. Many authors (Nayar & Chandra, 1965; Satija *et al.*, 1983) have distinguished several species such as *P. adnascens* (Sw.) Ching, *P. nuda* (Giesenh.) Ching and *P. varia* (Kaulf.) Farwell largely based on these narrow or otherwise flimsy characters. However, Hovenkamp (1986) reduced all these species under *P. lanceolata* as this taxon shows continuous variation. He has not recognized any infraspecific taxa under this species . In the present study, Hovenkamp's (1986) treatment has been followed.

Specimens examined : KERALA : Alappuzha Dt., Muhamma, *Sevichan* 29762; Pathiramanal, *Santhosh Nampy* 45178 (CALI). Kollam Dt., Thenmala, *Madhusoodanan* 49632 (CALI). Kottayam Dt., Vagamon, *Santhosh Nampy* 48866 (CALI). Kozhikode Dt., Arayadathupalam, *Santhosh Nampy* 48801 ; Mankavu, *Geevarghese* 4069; Vellimadukunnu, *Remani* 16558, *Rugmini Devi* 19518 (CALI). Malappuram Dt., Calicut Univ. Bot. Gard. (wild), *Santhosh Nampy* 44840; Chettiarmadu, *Santhosh Nampy* 44841; Kadampuzha, *Madhusoodanan* 7201 ; Thiruvangad, *Nayar* 17991 (CALI). Palakkad Dt., Cherunelly, *Nayar & Geevarghese* 7308; Dhoni hills, *Madhusoodanan* 21282; Kanjirapuzha, *Santhosh Nampy* 44838, *Geevarghese* 4016; Mannarghat, *Geevarghese* 4119; Nelliampathy, *Nayar* 17809; Silent Valley, *Geevarghese* 4137 (CALI). Wynad Dt., Adivaram, *Santhosh Nampy* 45950; Chandanathode, *Geevarghese* 4090, *Santhosh Nampy* 48887; Periya, *Santhosh Nampy* 45966; Sugandhagiri, *Santhosh Nampy* 45956; Vythiri, *Madhusoodanan* 4814, *Santhosh*

Nampy 48574 (CALI). TAMIL NADU : Coimbatore Dt., Valparai, *Santhosh Nampy & Madhusoodanan* 48871 (CALI).

Pyrrosia porosa (Presl) Hovenkamp, Blumea 30 : 208. 1984 & Leiden Bot. Ser. 9 : 226. 1986; Manickam & Irudayaraj, Pterid. Fl. W. Ghats 323. pl. 246. 1992; Nampy & Madhusoodanan, Indian Fern Jour. 9 : 214. 1992.

Type : India, *Wallich* Cat. no. 266 (holotype , PRC; isotypes, B, BM, P, US).

Polypodium porosum Wall., Cat. no. 266. 1828, *nom. nud.*

Niphobolus porosus Presl, Tent. Pterid. 200. 1836; Beddome, Ferns South India 61. pl. 183. 1864.

Cyclophorus porosus Presl, Epim. Bot. 130. 1851.

Niphobolus sticticus Kunze, Linnaea 24 : 257. 1851; Giesenh., Niphobolus 135. 1901.

Niphobolus fissus et syn. homot., auct. non Blume : Beddome, Handb. Ferns Brit. India 330. 1883.

Niphobolus gralla Giesenh., Niphobolus 128. 1901.

Pyrrosia gralla (Giesenh.) Ching, Bull. Chin. Bot. Soc. 1 : 50. 1935.

Pyrrosia mollis var. *mollisima* Ching, Bull. Chin. Bot. Soc. 1 : 54. 1935.

Niphobolus mollis et syn. homot., auct. non Kunze: Ching, Bull. Chin. Bot. Soc. 1 : 53. 1935; Sledge, Bull. Brit. Mus. (Nat. Hist.) Bot. 2 (5) : 134. 1960; Nayar & Chandra, Bull. Nat. Bot. Gard. Lucknow 117 : 67. 1965 & Can. Jour. Bot. 45: 615-634. 1967; Nayar & Kaur, Comp. Bedd. Handb. Ferns Brit. India 81. 1974; Devol & Kuo in Li *et al.* (eds.), Fl. Taiwan 1 : 207. 1975; Satija *et al.*, Bull. Bot. Surv. India 25 (1-4) : 76. 1983; Satija & Bir , Polyp. Ferns India 41. 1985; Chandra & Kaur, Nomen. Guide Bedd. Ferns South India & Ferns Brit. India 22. 1987.

Pyrrosia nayariana Ching & Chandra, Amer. Fern Jour. 54 : 62. t. 1-10. 1964; Nayar & Chandra, Bull. Nat. Bot. Gard. Lucknow 117 : 70. 1965 & Can. Jour. Bot. 45: 615-634. 1967; Satija *et al.*, Bull. Bot. Surv. India 25 (1-4) : 81. 1983; Satija & Bir, Polyp. Ferns India 49. 1985.

Pyrrosia stictica Holttum, Nov. Bot. Inst. Bot. Univ. Car. Prag. 31. 1968, *nom. superfl.*, Satija & Bir, Polyp. Ferns India 50. 1985; Manickam, Fern Fl. Palni Hills 151. 1986.

(Figs. 40g, 47)

Rhizome short creeping, 1.5-2 mm in diameter, densely paleaceous all over; cortical sclerenchyma sheath distinct, sclerenchyma strands numerous in the ground tissue, dictyostelic. Palea lanceolate, peltate, 4 x 0.75 mm, margin ciliate and paler, centre dark brown, apex long acuminate. Fronds monomorphic, closely placed, 3-6 mm apart, distinctly or indistinctly stipitate, simple, very occasionally the apex is abnormally forked; stipe not clearly demarcated being the lamina gradually narrowed and extending almost to the base of the frond, 0.2- 2 cm, base densely paleaceous, articulated; lamina oblanceolate or lanceolate, 9-32 x 0.8 - 3.4 cm, widest above or about the middle, thick, fleshy, brown below due to dense indumentum, glabrous adaxially, curl involutely on drying. Indumentum dimorphic, stellate, persistent; upper layer with straight-armed brown hairs, lower layer with long

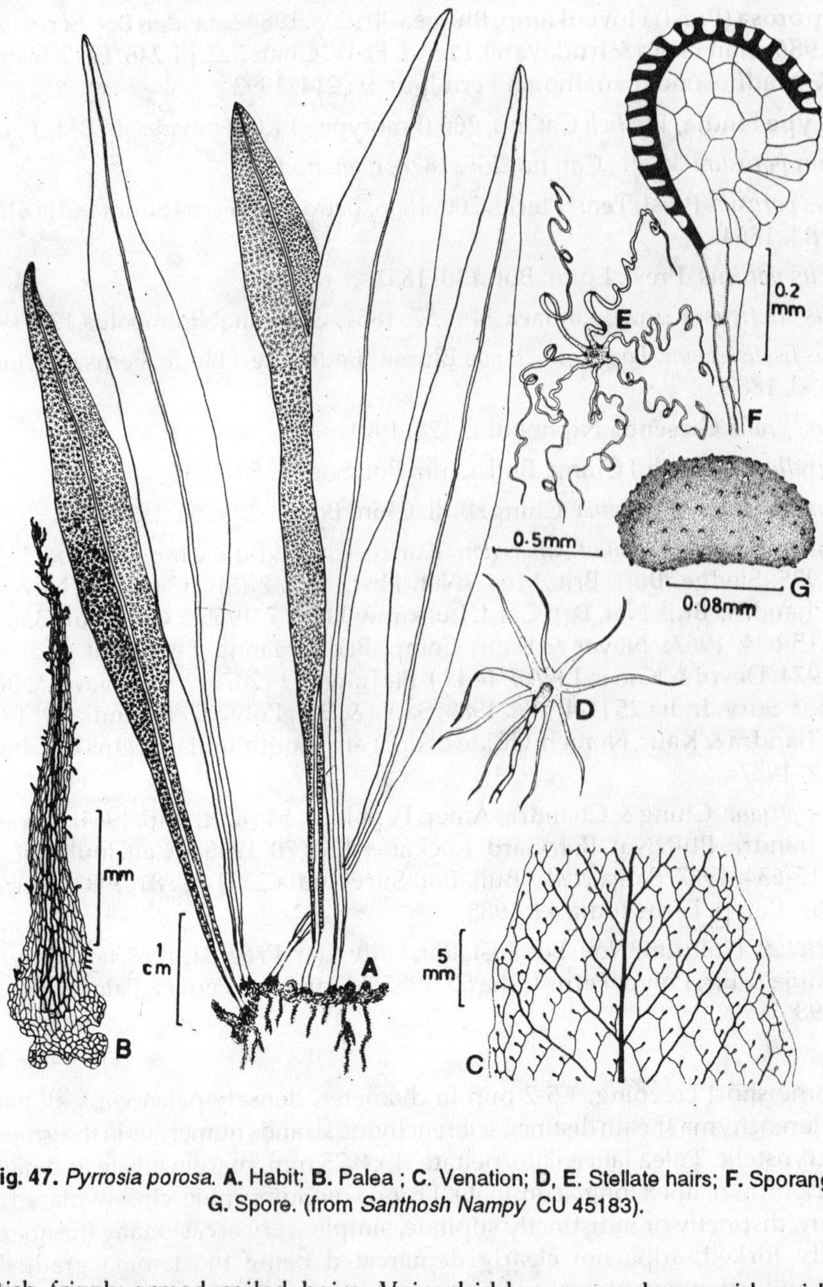

Fig. 47. *Pyrrosia porosa*. A. Habit; B. Palea ; C. Venation; D, E. Stellate hairs; F. Sporangium; G. Spore. (from *Santhosh Nampy* CU 45183).

whitish frizzly-armed coiled hairs. Veins hidden, compyloneuroid, midrib not raised on either side, lateral veins distinct, secondary veins form regular areoles with free excurrent veins, free excluded veins present. Sori apical or all over the lamina, at the end of included veins, distinct, closely packed, rarely confluent, covered by the indumentum. Sporangia stalked, stalk as long as the capsule, annulus 19 to 22-celled. Spores 70 x 52 µm, monolete, bilateral, light-yellow, sub verrucate.

Distribution and ecology : *P. porosa* is distributed in India, Burma, China, Java, Malesia and Philippines (Satija & Bir, 1985). It is a mountainous epiphyte or lithophyte, frequent on moist, moss covered rocks and trees in evergreen forests above 750 m (Fig. 43).

Notes : The South Indian specimens belonging to this species have been erroneously named either as *P. mollis* (Kunze) Ching (Sledge, 1960; Nayar & Chandra, 1965, 1967; Nayar & Kaur, 1974; Nayar & Geevarghese, 1993) or *P. stictica* (Kunze) Holttum (Satija *et al.*, 1983; Satija & Bir, 1985; Manickam, 1986). Hovenkamp (1986) correctly identified the South Indian specimens as *P. porosa* and this was followed in the subsequent treatments (Manickam & Irudayaraj, 1992; Nampy & Madhusoodanan, 1992). He recognised three varieties under it, *viz.*, var. *stenophylla* (Bedd.) Hovenkamp, var. *tonkinensis* (Giesenh.) Hoven kamp and var. *porosa* of which the latter is represented here. *P. porosa* can easily be distinguished from *P. mollis* by its dimorphic foliar indumentum and peltate palea with marginal hairs as opposed to monomorphic indumentum and pseudo-peltate, almost entire palea in *P. mollis* (currently known as *P. penangiana* (Hook.) Holttum). *P. penangiana* is reported to occur in South -east Asia and Pacific islands and there is no confirmed introduction of this species in India (Hovenkamp, 1986). *P. porosa* can be easily recognised from other South Indian members of this genus by its short creeping rhizome, larger fronds (9-32 x 0.8 - 3.4 cm) and dimorphic indumentum.

Specimens examined : KARNATAKA : Chikmagalore Dt., Bababudan hills, *Santhosh Nampy* 48843 (CALI). KERALA : Idukki Dt., Devikulam, *Santhosh Nampy* 48609; Pettimudy, *Madhusoodanan* 3834; Rajamala, *Madhusoodanan* & *Betzy* 1424 (CALI). Kozhikode Dt., Anakampoil, *Santhosh Nampy* 48857 (CALI). Malappuram Dt., Calicut Univ. campus, *Majeed* 20707; Thiruvangad, *Ramesh* 20499 (CALI). Palakkad Dt., Cherunelly, *Geevarghese* 7307, *Nayar & Geevarghese* 7309; Kaikatty, *Nayar* 7298; Nelliampathy, *Geevarghese* 17835; Panthenthode, *Nayar* 10074; Parambikulam, *Nayar* 7367, *Geevarghese* 17882; Silent Valley, *Nayar & Madhusoodanan* 21385, *Geevarghese* 4038; Vengoli peak, *Nayar & Pankajam* 7385 (CALI). Thiruvananthapuram Dt., Ponmudi, *Santhosh Nampy* 48880 (CALI). Wynad Dt., Chembra peak, *Nayar & Madhusoodanan* 21307, *Santhosh Nampy* 45915; Meppadi, *Madhusoodanan* 7275; *Pakshipathalam*, *Santhosh Nampy* 45973; Vythiri, *Nayar* & *Madhusoodanan* 7207 (CALI). TAMIL NADU: Coimbatore Dt., Valparai, *Santhosh Nampy* 48869; Yercaud, *Hema* 8227, *Usha* 30205 (CALI). Kattabomman Dt., Kothayar, *Madhusoodanan* 44887, *Santhosh Nampy* 45183 (CALI). Nilgiri Dt., Avalanchi, *Santhosh Nampy* 48805; Ooty, *Santhosh Nampy* 48803 (CALI). Quide Milleth Dt., Kodaikanal, *Sworrupanandan s. n.*, *Nayar* 2711 (CALI).

DISCUSSION

Comparative morphology

Polypodioid ferns, placed within the higher lepto-sporangiate ferns, constitute the largest group of extant ferns. Despite great diversity of form exhibited by its members, they are mainly confined to the tropical and warm temperate regions of the world. They are mostly epiphytic and invariably all members are destitute of indusium. Of the five families recognised within the polypodioid ferns by Wilson (1959), Nayar (1970) and Pichi Sermolli (1977), three are represented in South India. The families with respective number of genera and species in bracket are Grammitidaceae (3 genera and 8 species), Loxogrammaceae (1 genus and 2 species) and Polypodiaceae (10 genera and 23 species).

The South Indian polypodioid ferns are either epiphytic or epilithic and no terrestrial member has so far been recorded. A few, however, are adapted to live in aquatic habitats. Species like *Leptochilus metallicus, L. thwaitesianus* and *Microsorum pteropus* grow on shaded rock crevices beside water courses and can remain submerged during rainy seasons. *Leptochilus bahupunctika* grows in densely shaded forests among crowded undergrowth; it climbs up bases of woody shrubs and ultimately becomes semi-epiphytic as older parts of rhizome slough off.

Most of these ferns are restricted to undisturbed humid evergreen forests of medium to higher elevations (500-2600 m). A few like *Drynaria quercifolia, Pyrrosia heterophylla* and *P. lanceolata* are even adapted to live in dry exposed situations at low altitudes. Species like *Ctenopteris perplexa, Grammitis attenuata, Leptochilus metallicus, Loxogramme chinensis, Prosaptia contigua* and *P. obliquata* have a narrow distributional range and have been collected only from one or two places in South India, while other species of this group have comparatively more wider distribution in South India. The occurrence of *Colysis hemionitidea, Microsorum linguiforme, Phlebodium aureum* and *Phymatosorus scolopendria* in South India are considered to be an escape from cultivation. Moreover, all the above species are found to be naturalized in some places and is also widely grown as an ornamental in most of the botanical gardens and ferneries of South India.

Of the 33 species of South Indian polypodioid ferns 24 are found in Sri Lanka, suggesting a continuity in phytogeographic distribution, while only 17 species of this region are common to North India. Many species recorded earlier by Beddome

(1864-1892) like *Belvisia spicata, Prosaptia alata,* etc., were not able to trace out again even after extensive floristic survey in various parts of South India, suggesting that lack of proper conservational strategies may have led to the extinction of these species from this area. Since most of the polypodioid ferns are more precise in their habitat requirements only a few of them succeed under cultivation. Hence the *in situ* conservation in national parks and bioreserves is suggested. However, the new descriptions like *Grammitis pilifera, Leptochilus bahupunctika, Phymatosorus beddomei* and *P. malabaricus* show that intensive exploration in the still unattempted areas of Western Ghats can bring out many new taxa and at least some which are supposed to be extinct from this region.

The polypodioid ferns of South India are small or medium sized, having simple or dissected fronds. The smallest among them are species of *Grammitis,* which grow only upto 6 cm with simple, entire fronds. The larger forms like *Drynaria quercifolia, Phlebodium aureum* and *Phymatosorus nigrescens* have dissected fronds and usually grow upto 90 cm. Their rhizome is soft, fleshy, usually creeping either short or long, branched and sometimes forming clusters. The rhizome is thick and short creeping in *Microsorum punctatum* and *Pyrrosia porosa.* In the long creeping rhizomes, the fronds are distantly placed with long internodes. It is thin, slender and long creeping in *Pyrrosia ceylanica, P. heterophylla* and *P. lanceolata* or thick and long creeping in *Phlebodium aureum* and *Phymatosorus nigrescens.* A vertically climbing rhizome is seen in *Leptochilus bahupunctika* or it may be erect or sub-erect in *Grammitis pilifera* and *Ctenopteris perplexa.* The creeping or climbing rhizome is usually dorsiventral with a convex abaxial side and a more or less flat adaxial side, while the erect rhizome is usually terete. The diameter of the rhizome varies from 0.1 cm in *Grammitis* species to 3 cm in *Drynaria quercifolia.* The creeping rhizome is branched and branch buds usually alternate with fronds. A majority of branches remain dormant in most taxa. The colour of the rhizome is yellowish to light green when fresh turning brown or dark when dry. It is firmly attached to the substratum by roots situated adaxially. Roots are usually produced in two rows and covered with unicellular ferruginous hairs. The apex of the rhizome is densely paleaceous in most species while the older regions are sparsely so.

Fronds of all taxa (except Grammitidaceae and Loxogrammaceae, which are usually non-articulated) are functionally articulated at the base to phyllopodium. The articulation in most species being evident externally as a ring-like depression. The phyllopodium is quite prominent in some but insignificant in others, in drynarioid ferns the articulation at leaf base is least developed.

Vasculature of the rhizome of all Polypodiaceae and Loxogrammaceae is dictyostelic, being dissected into many slender, meristeles which form a loose cylindrical reticulum. In the strict sense of the term dictyostele (Schmid, 1982), the vascular cylinder of the rhizome of most Polypodiaceae is not dictyostelic since leaf gaps do not overlap. As common with other ferns, Polypodiaceae also possess a simple protostelic vascular cylinder in its juvenile plants, and from this there occurs a transition to the 'dictyostelic' adult condition. In Grammitidaceae, the vascular cylinder is either solenostelic or dictyostelic. In the primitive forms (*Grammitis)* the vasculature is solenostelic while in advanced forms (*Prosaptia),* it is dictyostelic.

Another important feature noticed in the rhizome anatomy of most of them is the presence of dark, longitudinally running sclerenchyma strands. The number of such strands may vary from one *(Pyrrosia ceylanica)* to many in others. However, such strands are absent in *Leptochilus thwaitesianus* and *Microsorum pteropus.*

Paleae form a dense felt on the growing region of rhizome in all species, however, they are less frequent in older regions. In a few genera like *Pleopeltis* and *Lepisorus,* paleae are also distributed on lamina which are persistent in the former. Glands are usually present at the apex and its margins, the glands are usually non-persistent. Palea are ovate-lanceolate in most species, however, it is lanceolate in *Ctenopteris perplexa,* oblong in *Grammitis* and lanceolate with a broadly ovate base gradually tapering into a long, slender, uniseriate hair-like structure in *Phymatopteris montana.* In *Pyrrosia lanceolata,* in addition to the lanceolate palea, shield-like ones are also occasionally present. They are peltate as in *Drynaria quercifolia* or basifixed as in *Loxogramme.* An important character used in defining taxa of this group is the grade of clathration of cell walls. They are usually clathrate or occasionally non-clathrate. The palea are uniformly pale-brown in *Lepisorus nudus, Phlebodium aureum* and species of *Loxogramme* or bicolorous with a dark central region and paler margins in *Microsorum membranaceum* and *Lepisours amaurolepidus.* In *Grammitis* species, the attaching region alone is having thick walled cells while other regions have thin walled cells. The palea may be entire *(Pyrrosia ceylanica, Lepisorus nudus)* or denticulate *(Pleopeltis macrocarpa)* or slightly dentate *(Colysis hemionitidea)* or ciliate *(Pyrrosia lanceolata).*

The fronds are borne in two alternating dorsal rows on the rhizome and may either be clustered as in *Microsorum punctatum* and *Pyrrosia porosa* or loosely placed as in *Microsorum linguiforme* and *Pyrrosia lanceolata.* They vary from simple to pinnatisect in different species. The fronds are simple *(Grammitis, Lepisorus, Leptochilus, Pyrrosia)* or pinnatisect *(Drynaria quercifolia, Phlebodium aureum).* In *Phymatosorus,* the pinnatifid condition is most common. The stipe may be prominent in most species or very rarely the fronds are sessile as in *Leptochilus metallicus.* A transition from simple to more complex pinnatifid condition can be seen in *Drynaria quercifolia* and *Phlebodium aureum.* In them the juvenile fronds are simple which become pinnatifid in the adult plant.

The rolling or wrinkling of lamina when dry is an important character in distinguishing species of *Loxogramme,* the frond margins roll revolutely in *L. chinensis* while they roll involutely in *L. cuspidata.* Similarly species of *Lepisorus, Pleopeltis* and *Pyrrosia* roll their lamina under dry habitats and can resurrect in wet conditions. The apex of lamina may be acute *(Pleopeltis macrocarpa)* or acuminate *(Microsorum linguiforme)* or blunt *(Leptochilus thwaitesianus).* Some species of this group have long decurrent base as in *Colysis hemionitidea* and *Leptochilus decurrens.* In *Microsorum linguiforme* occasionally the fronds become auricled at base. The fronds may be membranaceous as in *Microsorum membranaceum* or herbaceous as in *Lepisorus* species or coriaceous as in *Leptochilus metallicus.* The margins of lamina is often entire, but may be rather wavy in *Pleopeltis macrocarpa* or notched in *Phymatopteris montana.*

The fronds are glabrous in most species or rarely covered with indumentum of various kinds. In *Pleopeltis,* the fronds are moderately to densely set with persistent scales, while such laminar scales are deciduous in *Lepisorus.* Foliar appendages in *Pyrrosia* consist entirely of characteristic stellate hairs; palea and other appendages are absent on the leaf. The stellate hairs are found all over the lamina and stipe on young leaves but in most cases the hairs on the upper surface of the lamina are shed as the leaf matures. The nature and abundance of foliar hairs vary much among different species. In *P. porosa,* they are dimorphic forming an upper layer of broad armed stellate hairs and a lower layer of frizzly armed coiled hairs. In *P. lanceolata,* they are monomorphic consisting only of a thick layer of broad armed stellate hairs. Foliar hairs are less frequent in *P. heterophylla.* In *Grammitis,* the stipe and lamina are covered by reddish-brown acicular hairs similar to those found in Thelypteridaceae. In *G. pilifera,* few celled glandular hairs are also distributed abaxially.

The fronds in most species are monomorphic, while in some other genera, they are distinctly dimorphic. In *Leptochilus,* the vegetative fronds are simple, while the fertile lamina is highly contracted as a narrow 'wing' just enough to hold the sporangium. In *Drynaria,* the sterile fronds are sessile, broad, simple and humus collecting, while the fertile fronds are pinnatisect and prominently stipitate. In some species of *Pyrrosia* moderately to strongly dimorphic fronds are seen.

Venation is one of the important characters in defining genera of polypodioid ferns. It may be either free or anastomosing. Almost all species of Grammitidaceae have free venation, while in some species of *Ctenopteris,* the veins are reticulate. Very simple venation is seen in *Grammitis* where the lateral veins are simple or once forked but never they anastomose. In most species of *Ctenopteris,* the lateral veins remain free, however in a few species, the basal acroscopic veins are included. Similar venation pattern is also seen in other families like Gleicheniaceae and some species of Thelypteridaceae and Matoniaceae.

Members of Loxogrammaceae and Polypodiaceae have basically reticulate venation, with or without free included veins. In *Loxogramme,* the veins are free in the younger stage, but later they anastomose to form narrow oblique areoles usually without included veins. However, in some species (*L. cuspidata*), included veins are occasionally seen in some areoles. The vein endings in Loxogrammaceae are not swollen.

The venation of Polypodiaceae is the most diverse and complex within living ferns. The most complicated venation pattern found in Polypodiaceae is considered to have evolved with the simplicity of frond construction and the less complicated pattern of venation in *Polypodium* is usually considered to be a result of reduction (Copeland, 1947; Holttum, 1954). The vein endings in most species are swollen. The most simple type of venation in Polypodiaceae is that of *Polypodium vulgare,* where the veins are free in adult leaves (Mitsuta, 1983). The venation of *Lepisorus* and *Pleopeltis,* though reticulate, is rather simple with free included and excluded veins. The venation of *Phlebodium aureum* is peculiar and is known as *Venatio phlebodii.* The costal areoles are narrow without included veins, next a series of major areolae with excurrent included veins; beyond these, additional similar major areolae. The venation of *Pyrrosia* is also characteristic. Almost all species have reticulate venation with excurrent included veins. The venation may be compyloneuroid as in

P. lanceolata having included veins all excurrent and without included veins in the costal areoles or it may be microsorioid with simple or branched recurrent included veins as in *P. heterophylla.* The most complex and elaborate venation in Polypodiaceae is seen in drynarioid ferns. In this type of venation, the lateral veins are thick and straight, the secondary veins are distinct and rather straight, and within the areoles are many smaller areoles. These smaller areoles often lack included veins.

The sporangia are aggregated in naked sori borne superficially over the veins, usually at the vein pluxes, and are slightly or prominently immersed in depressions of the lamina. These depressed areas may protrude as pustules on the upper surface of the leaf lamina in *Phymatosorus* species. The sori are sunken in bilipped cavities in *Prosaptia contigua* and *P. obliquata.* The sori are either circular or oval in most species. They may be large and arranged in one row (*Lepisorus, Pleopeltis*) or more (*Phymatosorus scolopendria*), regular or irregular rows on either side of the midrib. It may be small and scattered over the lamina as in *Microsorum punctatum, M. membranaceum,* etc. In *Leptochilus,* the sori are acrostichoid and form continuous linear sori on the highly contracted fertile lamina. In *L. bahupunctika,* the fertile fronds are laminaless and the sori are directly placed on the midrib. In *Loxogramme,* the sori are elongate and oblique to the costa.

The sporangia are of the leptosporangiate type usually with a long stalk which is 3-celled at the top. However, in most grammitid ferns, the stalk is 1-rowed for most of its length. The annulus is 10-21 cells long and abuts on the stalk. The sporangia of Grammitidaceae bear one or a few stiff, reddish-brown acicular hairs similar to thelypteroid ferns. The paraphyses may be 2 or 3-celled as in *Microsorum* or long, filamentous as in *Phymatosorus beddomei* or with a 1-rowed stalk bearing two irregular rows of balloon-like cells ending in a swollen cell as in *P. nigrescens.* In *Pleopeltis,* the sori in young stage is protected by umbrella-shaped, dark brown paraphyses.

The spores are mostly monolete-bilateral but in all grammitid ferns, they are trilete-tetrahedral. In *Loxogramme,* the spores are either monolete or trilete or in some species both and intermediate spore types are seen on the same species (*L. chinensis).* The spores in most cases are perineless with smooth to variously ornamented exine. The exine is smooth in *Microsorum linguiforme, M. punctatum;* warty with conical spines in *Pyrrosia heterophylla;* spinulose in *Drynaria quercifolia* and coarsely tuberculate in *Phlebodium aureum.*

The three families of this group show a clear cytological affinity. In all of them, the numbers n=36 and n=37 have been counted by cytologists. Besides, n=35 has also been observed in the Loxogrammaceae and Polypodiaceae; while n=33 and other numbers occur occasionally in the Grammitidaceae and Polypodiaceae.

Three types of gametophytes have been observed in this group. The gametophyte of Grammitidaceae have a prolonged and extensive filamentous stage with a very slow growth rate. The adult prothallus is elongated and strap-shaped with a narrow median midrib bearing hairs of various kinds. The most common type of hair in this family is a branched, multicellular, club-shaped, glandular hair basically similar to the multicellular hair of the Polypodiaceae, but often more complex. Many species bear elongated, acicular hairs basically similar to those found in the Thelypteridaceae. The filamentous stage is very well defined in the Grammitidaceae and is considered as a specialized adaptation to the epiphytic habit.

The elongate, ribbon-like prothalli of Loxogrammaceae are much like those reported in some microsorioid genera of the Polypodiaceae (*Microsorum pteropus, Leptochilus, Colysis,* etc.). The thallus is slow-growing, perennial, thin, flat and without a midrib. The narrow, ribbon-like prothalli devoid of midrib and bearing sex organs on superficial marginal cushions are also found in some species of Hymenophyllaceae.

The gametophyte of Polypodiaceae is cordate-thalloid with a well differentiated meristem bearing glandular hairs and possessing distinct midrib or ribbon-like with prolonged growth, either hairy or sometimes naked, branched or devoid of a definite meristem or midrib. In some species of Polypodiaceae (*Lepisorus normalis*), the gametophyte is strap-shaped. Among the polypodioid ferns, the most primitive type is the cordate-thallus with a definite midrib and slow growth rate as seen in many genera of Polypodiaceae (*Pyrrosia*). The most advanced among them is the ribbon-like type occurring in many microsorioid genera (*Microsorum pteropus, Leptochilus* species) and *Loxogramme.*

Phylogeny

In the phyletic arrangement, we consider Grammitidaceae first followed by Loxogrammaceae and Polypodiaceae. Grammitidaceae is a homogenous group consisting of 12 genera. It has traditionally been regarded as a part of Polypodiaceae (Christensen, 1938; Copeland, 1947) or derived from them (Ching, 1940). The resemblance of Grammitidaceae to Thelypteridaceae as proposed by Nayar (1970) in their general morphology, non-articulated stipe with paired leaf traces, presence of acicular hairs on palea, frond and even on the prothallus appears to be superficial. When compared, it is found that the dissimilarities are more pronounced than similarities. The Grammitidaceae differs from Thelypteridaceae in their smaller size, epiphytic habit, exindusiate sorus, 1-rowed sporangial stalk, trilete spores and strap-shaped prothallus. This is in corroboration with Holttum's (1947) view that the two families differ so much so in other characters and that they are not so closely related, though both may originate from Gleicheniaceae. Nayar (1970) placed Grammitidaceae under Cyatheales while other polypodioid families under Gleicheniales. However, such an opinion appears to be unreliable since Grammitidaceae are bereft of indusium and seem out of place in a complex in which presence of a true indusium is almost a constant feature.

The Grammitidaceae is more close to Loxogrammaceae than any other group of modern ferns and have an origin from the gleicheniaceous stock along with ferns of other polypodioid group. Both these families share the following characters in common : their smaller size, epiphytic habit, non-articulate stipe, anomocytic stomata, exindusiate sorus, trilete spore and in the basic chromosome number. This is in corroboration with Pichi Sermolli (1977), who placed the Grammitidaceae under Polypodiales. Recently, the affinity of this family to Polypodiaceae (including *Loxogramme*) is also supported by Price (1983) and Wagner (1985).

The Loxogrammaceae is adopted here as an independent family with two genera, *Anarthropteris* and *Loxogramme.* Formerly, they were included under Polypodiaceae by several authors (Copeland, 1947; Holttum, 1947; Wagner, 1961; Pichi Sermolli, 1958). Loxogrammaceae occupies a position in between Grammitidaceae and Polypodiaceae and is more closer to the latter family in having dissected vascular cylinder, anastomosing venation, 3-rowed sporangial stalk and in the basic chromosome number, than to any other family. But it is sharply distinct from the latter in the non-articulate stipe and in the absence of anomocytic stomata. In the above characters, it shows affinity towards Grammitidaceae. Moreover, in the presence of monolete or trilete spores and ribbon-like prothallus, Loxogrammaceae appears to occupy an intermediate phylogenetical position between Grammitidaceae and Polypodiaceae.

The Polypodiaceae as construed here is a large and heterogenous family with more than 60 genera. Nayar (1970) considered it as the most advanced family among the homosporous ferns and proposed a gleicheniaceous origin. Pichi Sermolli (1977) though considered Polypodiaceae as belonging to the group of gleicheniaceous ferns, questioned the direct origin of them from Gleicheniaceae. He (*l.c.*), however, argued that they may be originated from ancestors allied to the Gleicheniaceae.

The origin of Polypodiaceae from a gleicheniaceous ancestor is found to be more realistic since Polypodiaceae and Gleicheniaceae share many characters in common. Almost every major morphological peculiarity of the sporophyte as well as the gametophyte of the Polypodiaceae can be traced back to the gleichenioid ferns. The peltate, gland-tipped palea, creeping rhizome, simple undissected lamina of the juvenile leaves, superficial sori, bilateral spores devoid of perine, Gleichenia-type of spore germination (in advanced genera of Polypodiaceae it is replaced by Vittaria-type), and Drynaria-type of prothallial development (in more advanced genera of Polypodiaceae, it is replaced by Kaulinia-type), etc. amply suggest a gleichenioid ancestory.

BIBLIOGRAPHY

Abraham, A. 1963. Cytology and phylogeny of the pteridophytes. *Mem. Indian Bot. Soc.* 4 : 1-7.

Abraham, A., C.A. Ninan and P.M. Mathew. 1962. Studies on the cytology and phylogeny of the pteridophytes. VII. Observations on one hundred species of South Indian ferns. *Jour. Indian Bot. Soc.* 41 : 339-421.

Agardh, J. G. 1839. *Recensio specierum generis pteridis.* Lundae.

Alston, A.H.G. 1956. The subdivision of the Polypodiaceae. *Taxon* 5 : 23-25.

Atkinson, L.R. and A.G. Stokey. 1964. Comparative morphology of the gametophyte of the homosporous ferns. *Phytomorph.* 14 : 51-70.

Baishya, A.K. and R.R. Rao. 1982. *Ferns and Fern Allies of Meghalaya State, India.* Jodhpur.

Balakrishnan, N.P. and S. Chowdhury. 1967. *Crypsinus chrysotrichus* (C. Chr.) Tagawa–A new fern record for India. *Sci. Cult.* 33 : 62-63.

Beddome, R.H. 1864. *The Ferns of Southern India.* Madras.

Beddome, R. H. 1866. *The Ferns of British India.* Madras.

Beddome, R.H. 1876. *Supplement to the Ferns of Southern India and British India.* Madras.

Beddome, R.H. 1883. *Handbook to the Ferns of British India, Ceylon and the Malay Peninsula.* Calcutta.

Beddome, R.H. 1892. *Supplement to the Ferns of British India, Ceylon and the Malay Peninsula.* Calcutta.

Bhargavan, P. 1973. Studies on South Indian ferns-III : Notes on two rare and little known taxa. *Bull. Bot. Surv. India* 15 : 281.

Bhargavan, P. and N.C. Nair. 1981. *Phlebodium aureum* (Linn.) J. Sm. (Polypodiaceae)–A new record for India. *Jour. Bombay Nat. Hist. Soc.* 77 : 539-540.

Bhavanandan, K.V. 1968. Studies on the cytology of sixteen species of South Indian ferns. *Caryologia* 21 : 333-338.

Bir, S.S. 1962. Taxonomy of Indian members of family Aspleniaceae. *Bull. Bot. Surv. India* 5 : 1-16.

Bir, S.S. 1965. Chromosome numbers in some ferns from Kodaikanal, South India. *Caryologia* 18 : 107-115.

Bir, S.S. 1973. Cytology of Indian pteridophytes. *In* P.K.K. Nair (ed.), *Glimpses in Plant Research.* Vol. I. New Delhi. pp. 28-119.

Bir, S.S. 1977. Pteridophytic flora of India : A review of achievements and future challenges in the systematics and taxonomy. *Bull. Bot. Surv. India* 19 : 323-329.

Bir, S.S. 1982. Classification and taxonomy of pteridophytes with particular reference to the higher ferns. *Aspects of Plant Sciences* 6 : 183-245.

Bir, S.S. and K. Devi. 1968. Taxonomic revision of the polypodiaceous genera of India–II. *Phymatodes* Presl. *Bull. Bot. Surv. India* 10 : 196-217.

Bir, S.S. and C.K. Satija. 1981. *Lepisorus kashyapii* in the Western Himalayas. *Amer. Fern Jour.* 71 : 53-56.

Bir, S.S. and P. Shukla. 1971. Pteridophytic flora of Simla hills : Families Loxogrammaceae and Polypodiaceae. *Nova Hedwigia* 21 : 193-224.

Bir, S.S. and C.K. Trikha. 1968a. Taxonomic revision of the polypodiaceous genera of India–1. *Microsorium* Link. *Bull. Bot. Surv. India* 10:133-148.

Bir, S.S. and C.K. Trikha. 1968b. Taxonomic revision of the polypodiaceous genera of India–III. *Pleopeltis. Amer. Fern Jour.* 58 : 119-125.

Bir, S.S. and C.K. Trikha. 1969. Taxonomic revision of the polypodiaceous genera of India-IV. *Polypodium lineare* complex and allied species. *Bull. Bot. Surv. India* 11 : 260-276.

Bir, S.S. and C.K. Trikha. 1974. Taxonomic revision of the polypodiaceous genera of India-VI. *Lepisorus excavatus* group. *Amer. Fern Jour.* 64(2) : 49-63.

Bir, S.S. and C.K. Trikha. 1979a. Cytological evolution of polypodioid ferns with particular reference to the Himalayan forms. *In* P.K.K. Nair (ed.), *Glimpses in Plant Research.* Vol. IV. New Delhi. pp. 98-130.

Bir, S.S. and C.K. Trikha. 1979b. The polypodioid ferns : their classification, characteristics and evolution. *New Botanist* 6: 25-32.

Bir, S.S., C.K. Trikha and S.M. Vasudeva. 1974. Taxonomic revision of the polypodiaceous genera of India-V. *Polypodium* L. and *Goniophlebium* (Bl.) Presl. *New Botanist* 1 : 142-159.

Bir, S.S., S.M. Vasudeva, P. Kachroo and Y. Singh. 1991. Pteridophytic flora of North-eastern India-V. Ecological, distributional and phytogeographical account. *Indian Fern Jour.* 8: 98–139.

Bishop, L.E. 1974. Revision of the genus *Adenophorus* (Grammitidaceae). *Brittonia* 26 : 217-240.

Bishop, L.E. 1977. The American species of *Grammitis* sect. *Grammitis. Amer. Fern Jour.* 67 : 101-106.

Bishop, L.E. 1978. Revision of the genus *Cochlidium* (Grammitidaceae). *Amer. Fern Jour.* 68 : 76-94.

Blûme, C.L. 1828. *Enumeratio Plantarum Javae et Insularum adjacentium. fasc. II. Filices.* Lugduni Batavorum.

Blume, C.L. 1829. *Florae Javae nec non Insularum adjacentium-Filices.* Bruxellis.

Bosman, M.T.M. 1986. Original *Microsorum* vs. later *Microsorium* (Polypodiaceae). *Taxon* 35 : 590-591.

Bosman, M.T.M. 1991. *A Monograph of the Fern Genus Microsorum* (Polypodiaceae). Leiden.

Brown, R. 1810. *Prodromus Florae Novae-Hollandiae et Insulae Van Diemen.* London.

Burmann, N.L. 1768. *Flora Indica.* Lugduni Batavorum.

Chandra, P. 1962. Ontogeny of peltate paleae in the Polypodiaceae. *Curr. Sci.* 31 : 477-479.

Chandra, S. 1982. Structure and organisation of the vascular system in the rhizome of drynarioid ferns. *Ann. Bot.* 50 : 585-598.

Chandra, S. and S. Kaur. 1987. *A Nomenclatural Guide to R.H. Beddome's Ferns of South India and Ferns of British India.* New Delhi.

Chiarugi, A. 1960. Tavole Chromosomiche delle Pteridophyta. *Caryologia* 13 : 27-150.

Ching, R.C. 1933a. The studies of Chinese ferns-IX. *Lepisorus* (J.Sm.) *gen. nov., Lemmaphyllum* Presl, *Neocheiropteris* Christ. *Bull. Fan Mem. Inst. Biol. (Bot.)* 4 : 47-113.

Ching, R.C. 1933b. The studies of Chinese ferns-X. *Microsorium, Colysis, Leptochilus, Selliguea. Bull. Fan Mem. Inst. Biol. (Bot.)* 4 : 293-352.

Ching, R.C. 1933c. A monograph of the fern genus *Phymatodes* Presl. *Contr. Inst. Bot. Nation Acad. Peiping* 2 : 55.

Ching, R. C. 1935. On the genus *Pyrrosia* Mirbel from the mainland of Asia including Japan and Formosa. *Bull. Chinese Bot. Soc.* 1 : 36-72.

Ching, R.C. 1940. On natural classification of the family 'Polypodiaceae'. *Sunyatsenia* 5 : 201-306.

Ching, R.C. 1964. On the genera *Phymatopsis* J. Sm. and *Crypsinus* Presl. *Acta Phytotax. Sin.* 9: 179-197.

Chowdhury, A. and G.S. Sarwade. 1982. A simple approach for climatic classification of India. *Tropical Ecology* 23 : 234-246.

Christensen, C. 1906. *Index Filicum.* Hafniae.

Christensen, C. 1931. Asiatic Pteridophyta collected by Joseph F. Rock 1920-1924. *Contr. U.S. Nat. Herb.* 26 : 265-337.

Christensen, C. 1934. *Index Filicum Supplementum Tertium.* Hafniae.

Christensen, C. 1938. Filicinae. *In* F. Verdoorn (ed.), *Manual of Pteridology.* The Hague. pp. 522-550.

Clarke, C.B. 1880. A review of the ferns of northern India. *Trans. Linn. Soc. Lond. (Bot.)* 1 : 425-611.

Copeland, E. B. 1928. *Leptochilus* and genera confused with it. *Philipp. Jour. Sci.* 37 : 333-415.

Copeland, E. B. 1929. The oriental genera of Polypodiaceae. *Univ. Calif. Publ. Bot.* 16 : 12-128.

Copeland, E.B. 1941. Comment on natural classification of the family Polypodiaceae by R.C. Ching. *Sunyatsenia* 6 : 159-177.

Copeland, E.B. 1947. *Genera Filicum.* New York.

Copeland, E.B. 1951. *Grammitis. Philipp. Jour. Sci.* 80 : 93-276.

Copeland, E.B. 1952. Grammitidaceae of New Guinea. *Philipp. Jour. Sci.* 81 : 81-119.

Copeland, E.B. 1955. *Ctenopteris* in America. *Philipp. Jour. Sci.* 84 : 381-470.

Copeland, E.B. 1960. *Fern Flora of the Philippines.* Manila.

Desvaux, A. N. 1811. Observations sur quelques nouveaux genres de fougeres. *Magaz. Gesellsch. Naturf. Freunde zu Berlin* 5 : 297-330.

Desvaux, A.N. 1827. Prodrome de la famille des fougeres. *Mem. Soc. Linn.* 6 : 171-337.

Devi, S. 1977. *Spores of Indian ferns.* New Delhi.

Devi, S. 1978. Studies on the spore morphology of some members of Polypodiaceae. *Brenesia* 14-15: 283-303.

Devi, S. 1981. *Reference Manual of Fern Spores.* Lucknow.

Devi, S. 1988. Spores of pteridophytes. *Indian Fern Jour.* 5: 28-57.

Devol, C.E. 1975. Grammitidaceae. *In* H. Li, T. Liu, T. Huang, T. Koyama and C.E. Devol (eds.), *Flora of Taiwan.* Vol. 1. Taipei. pp. 216-230.

Devol, C.E. and C.M. Kuo. 1975. Polypodiaceae. *In* H. Li, T. Liu, T. Huang, T. Koyama and C.E. Devol (eds.), *Flora of Taiwan.* Vol. 1. Taipei. pp. 165-214.

Diels, L. 1899. Polypodiaceae. *In* A. Engler and K. Prantl (eds.), *Die Naturlichen Pflanzenfamilien* 1 : 139-339.

Dixit, R.D. 1984. *A census of the Indian pteridophytes.* Howrah.

Dixit, R.D., D. Hazra and B.D. Kar. 1976. *Loxogramme remote-frondigera* (Hayata) C. Chr.—A new record for India and China. *Bull. Bot. Surv. India* 18 : 222-223.

Dixit, R.D. and J.N. Vohra. 1984. *A dictionary of the pteridophytes of India.* Howrah.

Erdtman, G. 1952. *Pollen Morphology and Plant Taxonomy.* Part-I. Stockholm.

Erdtman, G. 1957. *Pollen and Spore Morphology/Plant Taxonomy.* Part–II. *Gymnospermae, Pteridophyta, Bryophyta.* Stockholm.

Evans, A.M. 1963. New chromosome observations in the Polypodiaceae and Grammitidaceae. *Caryologia* 16 : 671-677.

Fabbri, F. 1957. Sondaggi Citogenetici nelle Polypodiaceae sensu stricto. *Caryologia* 10 : 402-407.

Fabbri, F. 1963. Primo Supplemento alle Tavole Chromosomiche delle Pteridophyta di Alberto Chiarugi. *Caryologia* 16: 235-335.

Fabbri, F. 1965. Secondo Supplemento alle Tavole Chromosomiche delle Pteridophyta di Alberto Chiarugi. *Caryologia* 18 : 675-731.

Farwell, O.A. 1931. Fern notes-II. *Amer. Midl. Naturalist* 12 : 233-311.

Fee, A.L.A. 1845. *Deuxieme memoire: Histoire des Acrostichees.* Strasbourg.

Fee, A.L.A. 1850-1852. Memoires sur les families des fougeres. Cinquieme memoire : Polypodiacees. *Mem. Soc. Hist. Nat.* Strasbourg 5 : 40-338.

Ghatak, J. 1963. Observations on the cytology and taxonomy of some ferns from India. *Proc. Indian Sci. Congr. Assoc. 55th Sess.* 3: 371-372.

Ghosh, S.R. 1985. A note on a species of fern genus *Phymatosorus* Pic. Serm. in India. *Jour. Econ. Tax. Bot.* 6 : 433-434.

Giesenhagen, K. 1901. *Die Farngattung Niphobolus.* Eine monographie. Jena.

Graham, J. 1836. *Catalogue of the Plants Growing in Bombay and its Vicinity.* Bombay.

Hartman, M.E. 1931. Antheridial dehiscence in the Polypodiaceae. *Bot. Gaz.* 91 : 252-276.

Hennipman, E. and M.C. Roos. 1982. A monograph of the fern genus *Platycerium* (Polypodiaceae). *Verh. Kon. Ned. Akad. Wet. Natuurk. Ser.* 2 : 1-126.

Hennipman, E., P. Veldhoen, K. U. Kramer and M.G. Price. 1991. Polypodiaceae. *In* K. Kubitzki (ed.), *The Families and Genera of Vascular Plants.* Vol. 1. New York. pp. 203-230.

Holttum, R.E. 1947. A revised classification of leptosporangiate ferns. *Bot. Jour. Linn. Soc.* 53 : 123-186.

Holttum, R.E. 1949. The classification of ferns. *Biol. Rev.* 24 : 267-296.

Holttum, R.E. 1954. A revised flora of Malaya–II. *Ferns of Malaya.* Singapore.

Hooker, W.J. 1842. *Genera Filicum.* London.

Hooker, W.J. 1837-1854. *Icones Plantarum.* London.

Hooker, W.J. 1846-1864. *Species Filicum.* London.

Hooker, W.J. and J.G. Baker. 1865-1868. *Synopsis Filicum.* London.

Hovenkamp, P. 1984. Some new names and combinations in *Pyrrosia* Mirbel (Polypodiaceae). *Blumea* 30 : 207-208.

Hovenkamp, P. 1986. *A Monograph of the Fern Genus Pyrrosia (Polypodiaceae).* Leiden.

Hozhizaki, B.J. 1981. The genus *Pyrrosia* in cultivation (Polypodiaceae). *Baileya* 21 : 53-76.

Ito, H. 1935. Nuntia ad Filices Japoniae III. *Jour. Jap. Bot.* 11 : 88-100.

Kaulfuss, G.F. 1824. *Enumeratio Filicum.* Lipsiae.

Kunze, G. 1846-1848. In Filices Javae Zollingerianas alias que ex herbario Morcandiano observationes. *Bot. Zeit.* 6 : 97-427.

Kunze, G. 1851. *Oleandra* Cav. species in herbario suo servatas sciagraphice disposuit. *Bot. Zeit.* 7: 1-345.

Kuriachan, P.I. 1965. Cytology of some Indian ferns. *Proc. Indian Sci. Cong. Assoc. 51st and 52nd Sess.* pp. 359.

Kuriachan, P.I. 1968. Cytological observations on some South Indian ferns. *Cytologia* 32 : 500-506.

Kuriachan, P.I. 1973. *Studies on the Cytology and Gametophytes of South Indian Pteridophytes.* Ph. D. thesis. Kerala University.

Labillardiere, J.J. H. de. 1806. Novae Hollandiae plantarum specimen. Parisiis.

Link, H.F. 1833. *Hortus regius botanicus berolinensis descriptus.* Berolini.

Linnaeus, C. 1753. *Species Plantarum.* Holmiae.

Lloyd, R.M. 1963. New chromosome numbers in *Polypodium L. Amer. Fern Jour.* 53 (3) : 99-101.

Lloyd, R.M. 1981. The perispore in *Polypodium* and related genera (Polypodiaceae). *Can. Jour. Bot.* 59 : 175-189.

Madhusoodanan, P.V. 1991. Rare and endangered ferns and fern allies of Western Ghats of Kerala. *Proc. In* K. Karunakaran (ed.), *Symp. Rare/Endangered/Endemic Plants of Western Ghats.* Thiruvananthapuram. pp. 103-107.

Madhusoodanan, P.V. and S. Nampy. 1993. The genus *Microsorum* Link in South India. *Jour. Econ. Tax. Bot.* 17 : 43-47.

Madhusoodanan, P.V. and S. Nampy. 1994. *Loxogramme chinensis* Ching, a misunderstood species from South India. *Indian Fern Jour.* 11 : 56-59.

Manickam, V.S. 1976. *Enumeration of Ferns of the Palni Hills.* Lucknow.

Manickam, V.S. 1984. Cytology of thirty species of ferns from Palni Hills–South India. *Cytologia* 49: 49-59.

Manickam, V.S. 1986. *Fern Flora of the Palni Hills (South India).* New Delhi.

Manickam, V.S. and V. Irudayaraj. 1988. *Cytology of Ferns of the Western Ghats-South India.* New Delhi.

Manickam, V.S. and V. Irudayaraj. 1992. *Pteridophyte Flora of the Western Ghats-South India.* New Delhi.

Manton, I. 1954. Cytological notes on one hundred species of Malayan Ferns. *In* R.E. Holttum (ed.), *Flora of Malaya.* Vol. 2. Singapore. pp. 623-627.

Manton, I. and W.A. Sledge. 1954. Observations on the cytology and taxonomy of Pteridophyte flora of Ceylon. *Phil. Trans. Roy. Soc.* 238: 127-185.

Meher-Homji, V.M. 1965. Ecological status of the montane grasslands of the South Indian hills. A phytogeographic reassessment. *Indian Forester* 91: 210-215.

Meher-Homji, V. M. 1967. Phytogeography of the South Indian hill stations. *Bull. Torrey Bot. Club* 94 : 230-242.

Mehra, P.N. and A. Gupta. 1986. *Gametophytes of Himalayan Ferns.* New Delhi.

Miquel, F.A.V. 1850-1852. Analecta Botanica Indica; seu Commentationes de variis stripibus Asiae Australioris. *Verhdl. Mederl. Inst.* 4 : 13.

Mirbel-Brisseau, C. 1803. *Histoire naturelle des vegetaux. In Deterville (ed.), Buffon : Cours complet d' Histoire naturelle.* Paris.

Mitsuta, S. 1983. Studies in the venation and systematics of Polypodiaceae-I. Venation (1). *Mem. Fac. Sci. Kyoto Univ. (Biol.).* 8 : 123-150.

Mitsuta, S. 1984a. Studies in the venation and systematics of Polypodiaceae–II. Venation of Polypodiaceae (2) and of some other ferns. *Mem. Fac. Sci. Kyoto Univ.* (Biol.) 9 : 57-85.

Mitsuta, S. 1984b. Studies in the venation and systematics of Polypodiaceae–III. Comments on generic classification of Polypodiaceae and notes on venation and systematics of Polypodiaceae. *Mem. Fac. Sci. Kyoto Univ. (Biol.)* 9 : 133-135.

Molly, M.J. 1984. *Organisation and Morphogenesis of the vascular system in Polypodiaceae (an inquiry into the nature of the shoot system in vascular plants).* Ph. D. thesis. Calicut University.

Moore, T. 1857. *Index Filicum.* London.

Morton, C.V. 1967. The Genus *Grammitis* in Ecuador. *Contr. U.S. Nat. Herb.* 38 : 85-123.

Morton, C.V. 1970. Recent fern literature : Reviews of tropical African Pteridophyta-I by E.A.C.L.E. Schelpe. *Amer. Fern Jour.* 60 : 123-127.

Nampy, S. and P.V. Madhusoodanan. 1992. Polypodioid ferns of South India. *Indian Fern Jour.* 9 : 204-220.

Nampy, S. and P.V. Madhusoodanan. 1995. *Loxogramme cuspidata* : A little known polypodioid fern from South India. *Jour. Econ. Tax. Bot.* 19(3) : 741-744.

Nayar, B.K. 1954. Contributions to the morphology of *Pseudodrynaria coronans. Phytomorph.* 3 : 411-423.

Nayar, B.K. 1955. Studies in Polypodiaceae–III. *Loxogramme. Jour. Indian Bot. Soc.* 34 : 396-407.

Nayar, B.K. 1957. Studies in Polypodiaceae–IV. *Drymoglossum. Jour. Indian Bot. Soc.* 36: 169-179.

Nayar, B.K. 1958. Studies in Polypodiaceae-VI. Further observations on the morphology of *Drynaria* Bory. *Jour. Univ. Gauhati* 9 : 95-103.

Nayar, B.K. 1959. Medicinal ferns of India. *Bull. Nat. Bot. Gard. Lucknow* 29 : 1-36.

Nayar, B.K. 1961a. Ferns of India-II. *Drynaria and Pseudodrynaria. Bull. Nat. Bot. Gard. Lucknow* 56: 1-30.

Nayar, B.K. 1961b. Ferns of India-III. *Microsorium. Bull. Nat. Bot. Gard. Lucknow* 58:1-38.

Nayar, B.K. 1961c. Studies in Polypodiaceae-VII. *Pyrrosia. Jour. Indian Bot. Soc.* 40 : 164-186.

Nayar, B.K. 1961d. Studies in Polypodiaceae-VIII. Morphology of the gametophyte of *Lepisorus excavatus* (Bory) Ching. *Sci. Cult.* 27: 345-347.

Nayar, B.K. 1962. Morphology of spores and prothalli of some species of Polypodiaceae. *Bot. Gaz.* 123 : 223-232.

Nayar, B.K. 1963a. Contributions to the morphology of *Leptochilus* and *Paraleptochilus. Amer. Jour. Bot.* 50: 301-308.

Nayar, B.K. 1963b. Contributions to the morphology of some species of *Microsorium. Ann. Bot.* 27 : 89-100.

Nayar, B.K. 1963c. Spore morphology of *Loxogramme. Grana Palynol.* 4 : 388-392.

Nayar, B.K. 1964. *Kaulinia*, a new genus of polypodiaceous ferns. *Taxon* 13 : 67-69.

Nayar, B.K. 1965. Gametophytes and juvenile leaves of drynarioid ferns. *Bot. Gaz.* 126: 46-52.

Nayar, B.K. 1968. The gametophyte and juvenile leaves of *Loxogramme. Amer. Fern Jour.* 58 : 19-29.

Nayar, B.K. 1970. A phylogenetic classification of the homosporous ferns. *Taxon* 19 : 229-236.

Nayar, B.K. and S. Chandra. 1965. Ferns of India-XV. *Pyrrosia* Mirbel. *Bull. Nat. Bot. Gard. Lucknow* 117: 1-98.

Nayar, B.K. and S. Chandra. 1967. Morphological series within the genus *Pyrrosia* and their phylogenetic interpretation. *Can. Jour. Bot.* 45 : 615-634.

Nayar, B.K. and S. Devi. 1964. Spore morphology of Indian ferns-III. Polypodiaceae. *Grana Palynol.* 5 : 342-395.

Nayar, B.K. and S. Devi. 1965. Spore morphology of Indian ferns-IV. Grammitidaceae. *Grana Palynol.* 6 : 121-127.

Nayar, B.K. and K.K. Geevarghese. 1993. *Fern Flora of Malabar.* New Delhi.

Nayar, B.K. and P. Kachroo. 1953. Studies in Polypodiaceae–I. Contributions to the morphology of *Drynaria* Bory: *D. quercifolia* (L.) J. Sm. and *D. propinqua* (Wall.) J. Sm. *Phytomorph.* 3 : 411-423.

Nayar, B.K. and S. Kaur. 1971. Gametophytes of homosporous ferns. *Bot. Rev.* 37 : 295-396.

Nayar, B. K. and S. Kaur. 1974. *Companion to Beddome's Handbook to the Ferns of British India.* New Delhi.

Nayar, B.K. and P.V. Madhusoodanan. 1977. Some aspects of the morphology and distribution of *Microsorium linguaeforme* (Polypodiaceae). *Bot. Jour. Linn. Soc.* 75 : 283-298.

Nayar, B.K., P.V. Madhusoodanan and M.J. Molly. 1985. *Nistarika*, a new genus of Polypodiaceae from Silent Valley, South India. *Fern Gaz.* 13 : 33-42.

Nayar, B.K. and F. Raza 1970. The prothalli of some Polypodiaceae–II. *Jour. Indian Bot. Soc.* 49 : 81-86.

Newman, E. 1840. *A History of British Ferns and Allied Plants.* London.

Pal, S. and N. Pal. 1963. A new species of *Phymatodes* in the *Phymatodes scolopendria* Complex. *Amer. Fern Jour.* 53 : 101-109.

Parris, B.S. 1975. A revision of the genus *Grammitis* Swartz (Filicales : Grammitidaceae) in Australia. *Bot. Jour. Linn. Soc.* 70: 21-43.

Parris, B.S. 1983. A taxonomic revision of the genus *Grammitis* Swartz (Grammitidaceae : Filicales) in New Guinea. *Blumea* 29 : 13-222.

Parris, B.S. 1991. Grammitidaceae. *In* K. Kubitzki (ed.), *The Families and Genera of Vascular Plants.* Vol. 1. New York. pp. 153-157.

Parris, B.S. and D.R. Given. 1976. A taxonomic revision of the genus *Grammitis* Swartz (Grammitidaceae: Filicales) in New Zealand. *New Zeal. Jour. Bot.* 14 : 85-111.

Patel. J.D., E.C. Raju, R.L. Fotedar, I.L. Kothari and J.J. Shah. 1975. Structure and histochemistry of stomata and epidermal cells in five species of Polypodiaceae. *Ann. Bot.* 38 : 611-619.

Pichi Sermolli, R.E.G. 1953. The nomenclature of some fern genera. *Webbia* 9 : 387-454.

Pichi Sermolli, R.E.G. 1958. The higher taxa of the Pteridophyta and their classification. *In* O. Hedberg (ed.), *Systematics of Today.* Uppsala Univ. Arrskr. pp. 70-90.

Pichi Sermolli, R.E.G. 1959. Pteridophyta. *In* W.B. Turrill (ed.), *Vistas in Botany.* London. pp. 421-493.

Pichi Sermolli, R.E.G. 1965. The nomenclature of *Pleopeltis lanceolata. Webbia* 20 : 349-354.

Pichi Sermolli, R.E.G. 1973. Fragmenta Pteridologiae-IV. *Webbia* 28: 445-477.

Pichi Sermolli, R.E.G. 1974. Fragmenta Pteridologiae-V. *Webbia* 29 : 1-16.

Pichi Sermolli, R.E.G. 1977. Tentamen Pteridophytorum genera in taxonomicum ordinem redigendi. *Webbia* 31 : 313-512.

Pichi Sermolli, R.E.G. 1982. A further contribution to the nomenclature of the families of Pteridophyta. *Webbia* 35 : 223-237.

Presl, C.B. 1836. *Tentamen Pteridographiae.* Prague.

Price, M.G. 1983. *Pecluma*, a new tropical American fern genus. *Amer. Fern Jour.* 73 : 109-116.

Price, M.G. 1984. Three new combinations in *Loxogramme. Amer. Fern Jour.* 74: 61.

Price, M.G. 1990. Four new Asian *Loxogramme. Amer. Fern Jour.* 80: 4-8.

Price, M.G. 1991. *Loxogramme. In* K. Kubitzki (ed.), *The Families and Genera of Vascular Plants.* Vol. 1. New York. pp. 229.

Proctor, G.R. 1977. *Flora of the Lesser Antilles.* Vol. 2. *Pteridophyta.* Cambridge.

Rao, K.N., C.J. George and K.S. Ramasastry. 1972. *Agroclimatic Classification of India.* India.

Ravi, N. and J. Joseph. 1979. A new species of fern from South India. *Jour. Bombay Nat. Hist. Soc.* 76 : 348-350.

Rheede, H.A. van. 1678-1703. *Hortus Indicus Malabaricus.* Amsterdam.

Rodl-Linder, G. 1990. A monograph of the fern genus *Goniophlebium* (Polypodiaceae). *Blumea* 34 : 277-423.

Roos, M. C. 1985. *Phylogenetic systematics of the Drynarioideae (Polypodiaceae).* Universiteitsdrukkerij, Utrecht.

Roth, A.W. 1821. *Novae Plantarum Species Praeserti Indiae Orientalis.* Halberstadii.

Sankari Ammal, L. and K.V. Bhavanandan. 1991. Studies on the cytology of some members of Polypodiaceae (sensu Copeland). *Indian Fern Jour.* 8: 154-158.

Satija, C.K. and S.S. Bir. 1985. *Polypodiaceous Ferns of India.* New Delhi.

Satija, C.K., S.S. Bir and A.K. Bharadwaj. 1983. Taxonomic revision of the polypodiaceous genera of India. *Pyrrosia* Mirbel. *Bull. Bot. Surv. India* 25 : 62-89.

Schelpe, E.A.C.L.E. 1970. *Flora Zambesiaca-Pteridophyta.* London.

Schmid, R. 1982. The terminology and classification of steles : historical perspective and the outlines of a system. *Bot. Rev.* 48 : 817-931.

Shing, K.H. 1983. A reclassification of the fern genus *Pyrrosia. Amer. Fern Jour.* 73 : 73-78.

Sledge, W.A. 1956. The Ceylon species of *Leptochilus. Ann. & Mag. Nat. Hist.* 12 : 865-877.

Sledge, W.A. 1960. The Polypodiaceae and Grammitidaceae of Ceylon. *Bull. Brit. Mus. (Nat. Hist.) Bot.* 2: 133-157.

Sledge, W.A. 1967. *Grammitis* in Samoa. *Fern Gaz.* 9 : 339-347.

Smith, J. 1841. Enumeration Filicum Philippinarum or a systematic arrangement of the ferns collected by H. Cuming Esq. in the Philippine islands and the Peninsula of Malacca between the years 1836 and 1840. *Jour. Bot.* 3 : 392-422.

Smith, J. 1842. An arrangement and definition of the genera of ferns with observation on the affinities of each genus. *Jour. Bot.* 4 : 38-70.

Smith, J. 1846. An enumeration of ferns cultivated in the Royal Botanic Gardens at Kew in December 1845, with characters and observations on some of the genera and species. *Comp. Bot. Mag.*

Smith, J. 1875. *Historia Filicum.* London.

Sota, E.R. de la. 1973. On the classification and phylogeny of the Polypodiaceae. *In* A.C. Jermy, J.A. Crabbe and B.A. Thomas (eds.), *The Phylogeny and Classification of the Ferns.* London. pp. 229-244.

Stone, I.G. 1960. Observations on the gametophytes of *Grammitis billardieri* Willd. and *Ctenopteris heterophylla* (Labill.) Tindale (Grammitidaceae). *Austr. Jour. Bot.* 8 : 11-37.

Subramanyam, V.P, S.B. Rao and A.R. Subramanyam. 1965. Koppen and Thornth waite system of climatic classification as applied to India. *Ann. Arid Zone* 4: 46-55.

Swartz, O. 1801. Genera and species filicum ordine systematico redactarum. *Schrader's Jour. Bot.* 1800 : 1-120.

Swartz, O. 1806. *Synopsis Filicum, earum genera et species systematice complectens.* Kiliae.

Tryon, R.M. and A.F. Tryon. 1982. *Ferns and Allied Plants with Special Reference to Tropical America.* New York.

Uffelen, G.A. van. 1993. Sporogenesis in Polypodiaceae (Filicales)–III. Species of several genera. Spore characters and their value in phylogenetic analysis. *Blumea* 37 (2) : 529-562.

Underwood, L.N. 1903. A summary of our present knowledge of the ferns of the Philippines. *Bull. Torr. Bot. Club.* 30 : 665-683.

Vasudeva, S.M. and S.S. Bir. 1983. Chromosome numbers and evolutionary status of ferns and fern allies of Pachmahri hills (central India). *Intern. Biosci. Monograph-13.* New Delhi.

Wagner, F.S. 1985. Bilateral spores in New World grammitid ferns. *Amer. Fern Jour.* 75 : 6-11.

Wagner, W.H. 1961. Problems in the classification of ferns. *Recent advances in Botany.* Toronto. pp. 841-844.

Willdenow, K.L. 1810. *Carolia Linne Species Plantarum.* Berolini.

Wilson, K.A. 1958. Ontogeny of the sporangia in *Xiphopteris serrulata* and *Pyrrosia nuda. Jour. Arn. Arbor.* 39 : 478-493.

Wilson, K.A. 1959. Sporangia of the fern genera allied with *Polypodium* and *Vittaria. Contr. Gray Herb. Harv.* 185: 97-127.

Zenker, J.C. 1835. *Plantae Indicae quas montibus Coimbaturicis coeruleis Nilagiri sive Neilgherries dictis collegit.* Rev. Bernh. Schmid. Jena. Fol.

Zink, M.J. 1988. Affinities and delimitation of the genus *Lepisorus* (J. Sm.) Ching (Polypodiaceae). *In* K.H. Shing and K.U. Kramer (eds.), *Proc. Intern. Symp. Systematic Pteridology.* Beijing. pp. 75-81.

INDEX TO SCIENTIFIC NAMES